D1172479

EFFECTIVE EXPERT WITNESSING

Second Edition

JACK V. MATSON

LEWIS PUBLISHERS
Boca Raton   Ann Arbor   London   Tokyo

SEP 0 6 1995

KF
8961
.M38
1994

Library of Congress Cataloging-in-Publication Data

Matson, Jack V.
    Effective expert witnessing / Jack V. Matson. — 2nd ed.
        p.   cm.
    Includes bibliographical references and index.
    ISBN 1-56670-002-7
    1. Evidence, Expert—United States.   I. Title.
    KF8961.M38   1994
    347.73'67—dc20
    [347.30767]   94-15811
                                                                                    CIP

This book contains information obtained from authentic and highly regarded sources. Re-printed material is quoted with permission, and sources are indicated. A wide variety of references are listed. Reasonable efforts have been made to publish reliable data and information, but the author and the publisher cannot assume responsibility for the validity of all materials or for the consequences of their use.

Neither this book nor any part may be reproduced or transmitted in any form or by any means, electronic or mechanical, including photocopying, microfilming, and recording, or by any infor-mation storage or retrieval system, without prior permission in writing from the publisher.

CRC Press, Inc.'s consent does not extend to copying for general distribution, for promotion, for creating new works, or for resale. Specific permission must be obtained in writing from CRC Press for such copying.

Direct all inquiries to CRC Press, Inc., 2000 Corporate Blvd., N.W., Boca Raton, Florida 33431.

© 1994 by CRC Press, Inc.
Lewis Publishers is an imprint of CRC Press

No claim to original U.S. Government works
International Standard Book Number 1-56670-002-7
Library of Congress Card Number 94-15811
Printed in the United States of America 1 2 3 4 5 6 7 8 9 0
Printed on acid-free paper

OCLC # 30318644

SEP 0 6 1995

Preface

The few books on the subject of expert witnessing either are by lawyers on how to destroy experts, or they are limited to descriptions of the duties of the expert with only a brief mention of the legal process. The purpose of this book is to get into the "head-to-head" battles that characterize legal warfare in a courtroom. An expert needs to fight the intellectual battles with knowledge of the tactics and strategies employed and how to be effective. Through this book, I want to familiarize you with the mind games lawyers play and ways to plot out your own battle plans.

As with any other project, you need to prepare for expert witnessing by doing research. This book will not only give you background, but will also propel you into the fascinating, complex, and somewhat awesome world where technology and law intersect.

The material for the book grew from a syllabus I prepared for a series of short courses on expert witnessing. Jon Lewis had approached me about writing a book for environmental professionals, but when I showed him the expert witnessing syllabus, he was excited about its potential. This is the second edition. I decided to use six of my cases to illustrate the narrative material, plus a chapter on the future of expert witnessing. The cases were selected to bring out the most important facets of witnessing in real fact situations. The names of experts and litigants have been changed, and the fact situations simplified to allow you, the expert, to vicariously live through these cases with me and to understand my reasoning, what I did, the problems I faced, and the insights I gained.

Section I deals with how to be an expert witness. It includes the following chapters. Chapter 1 confronts you with the dilemmas of the expert witness; Chapter 2 provides necessary factual and procedural information about our court system and what you need to know as an expert. Chapter 3 outlines what is expected of you — by your attorney and by the legal system — as an expert. Chapter 4 gives specific guidance on marketing your expert skills, while Chapter 5 describes the all important phases of litigation. Chapter 6 takes you into the courtroom offering detailed information on direct and cross-examination, the most critical area of your duties as an expert. Chapter 7 discusses feelings, emotions, and other intangible elements of testimony. In Section II, seven case studies are presented. Chapter 8 is a simulated case *in toto*, the litigation process in microcosm. Chapters 9 through 14 are the illustrated cases. In Section III, further observations and conclusions are drawn. Chapter 15 summarizes the underlying psychological challenges of litigation for the expert witness and the attorneys. Chapter 16 deals with the verdict, and Chapter 17 addresses the future of expert witnessing, offering a few concluding thoughts covering subtle but important aspects of expert witnessing.

Throughout the book, question-and-answer examples of courtroom testimony are presented, then dissected and analyzed. (Please note that where I refer to the actors in the masculine "he", the feminine "she" is just as appropriate.) The examples come from real life experiences. I'd like to help you to learn from my mistakes and my successes and to assist you with the legal system, so that your expert knowledge can shine through the legal smoke. You can be of great value.

The legal system sorely needs competent, qualified experts who can deal with technical complexities in the courtroom. Being an expert witness in a court of law can be the most demanding, challenging, and exhilarating experience of your career.

I wish to thank first and foremost Elizabeth Goreham who helped launch the book and prepare the second edition. The attorneys who not only retained me in their important cases but helped teach the expert witness short courses with me and reviewed the book were Jim Blackburn, Denise Morris Hammond, and Debra Barnhart Todd. They became and remain good friends.

About The Author

Jack V. Matson is the Director of the Leonhard Center for Innovation and the Enhancement of Engineering Education, and Professor of Civil and Environmental Engineering at the Pennsylvania State University. He is an expert in waste management, industrial water and wastewater treatment, hazardous waste, and air pollution; and has authored over 50 publications, two patents, and another book: *The Art of Innovation Using Intelligent Fast Failure*. Dr. Matson develops courses and curricula in innovative engineering design to encourage teamwork and creative problem solving in students, as well as teaches and does research in environmental engineering.

Dr. Matson received BS and MS degrees in Chemical Engineering from the University of Toledo and a Ph.D. in Environmental Engineering from Rice University. He also attended the University of Michigan Law School.

Before entering the academic realm, Dr. Matson was a Process Chemical Engineer for the Sun Oil Refinery, Toledo, Ohio; Environmental Engineer for the Enjay (now Exxon) Chemical Company in Baytown, Texas; and Manager of Environmental Engineering for S&B Engineers and Contractors in Houston, Texas. As a consultant, Dr. Matson has participated in the design and construction of numerous waste treatment facilities.

Starting in the mid-1970s, Dr. Matson began giving testimony as an expert at State of Texas administrative hearings involving environmental permits. Since then, he has participated in a variety of significant cases involving his area of expertise.

As a member of the Texas Air Control Board, he helped set policy on the implementation of the 1990 Federal Clean Air Act. Also, he continues to give short courses on expert witnessing. His address is: 202 Rider Building II, Penn State University, University Park, Pennsylvania, 16802; phone 814: 865-4014.

Contents

SECTION III: OBSERVATIONS AND CONCLUSIONS

Section I

How to Be an Expert

1

Baptism of the Expert Witness

Bailiff: Do you solemnly swear to tell the truth and nothing but the truth, so help you God?

"What am I doing here on the witness stand? I'm a very successful clothier and here I am testifying as an Expert against a tailor. Imagine me, an Expert witness."

FACTS OF THE CASE

An individual rushed into a tailor's shop and requested that the tailor make a suit based on measurements the customer had had taken earlier. After a discussion in which the tailor initially refused the job, he was persuaded to go ahead with the suit. Later, when the suit did not fit, the customer sued the tailor to get his money back. The tailor is being sued for negligence on the basis that it is the duty of a tailor to accept only his own measurements or those of another qualified tailor. He breached that duty by accepting the measurements of the client. The tailor is claiming that the customer waived the tailor's duty by assuming the risk after the tailor's initial refusal.

Q: Sir, what is your background and experience?

A: After high school I went to the State University and received a degree in Textiles. I worked for Zell Brothers

Clothes for 15 years doing everything from tailoring to buying and retailing. Five years ago I started my own store and now have a chain of ten.

Q: Have you written any articles?

A: Yes, I've written twelve articles and a book entitled *How to Tailor-Make Your Own Business.*

Q: How would you explain what happened in this case?

A: When the customer demanded a custom-made suit using measurements provided by the customer, the tailor should have refused. No tailor should ever accept measurements unless he has confidence in them. Even then the tailor must be careful. His reputation is at stake. I never accept measurements if they are taken by another tailor.

Q: Do you have an opinion as to the duty of the tailor in this instance?

A: Yes.

Q: What is your opinion?

A: The tailor breached his duty and was negligent in accepting the customer's measurements.

"Oh-oh. Here is another attorney. This must be cross examination. I wish my lawyer had prepared me for this."

Q: Sir, the tailor initially refused to take the order, didn't he?

A: Yes.

Q: And he refused because he wanted to take the measurements, didn't he?

A: Yes.

Q: So the customer was put on notice that the measurements were crucial, correct?

A: Yes.

Q: The customer insisted that the tailor take the order, using the measurements he provided, didn't he?

A: Yes.

Q: Have you ever had that happen to you, where the customer insisted on providing his own measurements?

A: No.

Q: Thank you. No further questions.

"What an empty feeling. What could I do? The lawyer trapped me with those questions. I had no choice but to agree with him. Or was there a better way to answer?"

The dilemma of this Expert is obvious. He wanted to answer in a truthful, forthright manner. But in doing so, he was trapped by the examiner into admissions that were devastating.

How could he have answered just as truthfully, without yielding damaging admissions to the other side? After reading this book, you will know.

2

Interfacing with the Legal System

The following chapter is designed to be a ready reference to legal procedure, terminology, and sequence of events in a civil case. All information is from the perspective of the Expert, so that he or she may be properly prepared at every stage.

PRETRIAL DISCOVERY

Discovery Process

The legal system provides a mechanism for the discovery of evidence in a lawsuit prior to the trial. The strategy is to push both sides toward settlement, if possible. And it does away with trial by ambush. Both the Plaintiff and the Defendant are entitled to the fullest disclosure of knowledge pertinent to the case, subject to certain limitations such as privilege and the attorney work product doctrine.

Pleadings and Motions

In most cases, the first formal notice of a lawsuit is the Complaint. It is a legal document written in very general terms alleging some cause for action, i.e., the harm that has been done to a party. An example would be a breach of contract in which one party did not perform or do agreed upon items. Another cause for action is negligence, in which one party violates the standard of care in the fulfillment of an obligation.

Provided there are no objections to the complaint, the opposing party, who receives the Complaint, files an Answer. In a typical answer all claims are denied. So called Affirmative Defenses are presented. The most common affirmative defenses are waiver, estoppel, and third party claims. In the waiver defense, the claim is that the plaintiff in some way gave up the right to sue. For example, they assumed the risk: a piece of machinery was not working; instead of contacting the manufacturer, they attempted the repair. The defendant will claim the plaintiff waived the right to sue as stated in the warranty.

In estoppel, the plaintiff is "stopped" from suing if they are claiming something contrary to what has been established as the truth. For example, if the plaintiff originally agreed by valid contract to construct a bridge, they are estopped from denying such a contract existed.

Third party claims involve shifting the responsibility to another party. The defendant states that they did not do it, that "so and so" did. For example, it could be a subcontractor that is responsible. For all affirmative defenses, the burden of proof resides with the defendant.

The Expert can be helpful to the attorney in constructing the Complaint, or the Answer, for their side. They can identify other parties who should be sued (third party claims and cross claims), and identify the various responsibilities of the parties involved.

Documents and Tangible Evidence

Both parties file a "Request for Production," which asks for all records on the particular project being litigated. The attorneys can request work papers, memos, letters, sales brochures, financial data, résumés, production records, etc.

You need to educate your lawyer on the types of documents that are available. For example, a service company may have

filed periodic reports on the status of some equipment or operation that may contain useful information.

What is not subject to production is defined in three primary
privileges:

1. *Attorney-Client Privilege.* Communications between clients
   and their attorneys are protected under this privilege.

2. *Attorney Work Product.* Documents prepared in anticipa-
   tion of litigation under the supervision of or at the direction of an attorney are not discoverable. Working papers
   prepared under the direction of an attorney, such as notes
   of telephone calls, are also exempt from discovery. But
   be careful what you write. Some work product documents
   can be discovered, such as calculations that will be introduced as evidence. Also, there are times when the judge
   will require you to produce documents if the information
   is no longer available from another source. You may have
   the only copy of an important drawing, for example.

3. *Proprietary Processes and Patents.* Information that is vul-
   nerable to exposure to competitors may be exempt. Or, it
   may be subject to production under a protective order
   issued by the court. The Expert can be of great assistance
   to the lawyer in understanding what is and is not proprietary.

Questions about privilege are dealt with by the judge "in
camera" (in the judge's chambers). The judge reviews the material in secret and makes a ruling.

Freedom of Information Acts

The federal government, states, and local governments must
make all records available to any citizen. Generally, you must
make the request in writing and pay for the reproduction costs.
Valuable information can be gleaned from these files if the
parties to the lawsuit dealt with public entities.

Expert Reports

There are two kinds of Experts—consulting and testifying. The consulting Expert provides background knowledge. Under some states' rules, none of their work is discoverable except reports or affidavits prepared for use at trial. The testifying Expert retains work product privileges except for papers used at trial.

Do not write any reports unless instructed to do so by your lawyer. The worst thing that can happen is for your attorney to receive a report that challenges the case, or a phone conversation is converted to a report that is discovered.

Information you relied on to develop your opinion is discoverable. Papers, articles, memos, calculations, and facts you used to prepare your expert report must be made available to the other side.

Interrogatories

Interrogatories are written questions sent to the other side. The recipient is given thirty days to provide written answers. It is not easy to construct or answer interrogatories. Help your lawyer develop questions for certain information, such as:

1. *Witness Information*: Who was there? Who saw what happened? Who had the authority to act? Who was involved?

2. *Remedial Action*: What did they do and why did they do it? Dates, times, places, etc.

Depositions

Lawyers for both sides can notice or subpoena anyone associated with the case to give oral testimony under oath. In some states, this includes the Expert Witness. Although no judge is present, the witness is sworn to tell the truth by a court reporter and a transcript of the testimony is prepared.

The Expert can be of great assistance to the lawyer by helping to prepare lines of questioning and analyzing the other side's case: Where are they coming from? Will they be saying the case is a misapplication, misdesign, or misoperation? The Expert can help analyze technical theories and lines of questioning.

The Expert can be, and often is, present at the deposition. He or she can study the personalities involved, recognize when the witness is weaseling, and figure out how to pin them down. You can write questions on a piece of paper and pass it along to the lawyer or even whisper in their ear. During breaks you can step outside and discuss the situation.

Depositions provide the best, most direct way to gather intelligence. The deposition transcript becomes a written record upon which witnesses can be impeached at trial if their testimony varies from the deposition statement.

Subpoenas

A subpoena can be used to compel you (or any witness) to appear at a deposition or trial. To be valid, the subpoena must be served on you personally by a constable or a process server. A simple subpoena is a written legal document directing you to appear at a particular time and place to testify as a witness. A subpoena *duces tecum* not only requires your attendance and testimony, but also requires that you bring along specified materials, for example, your files.

If you are subpoenaed, you cannot decline to testify. However, for a deposition, lawyers will in general allow you to specify the time and place at your convenience. You can only be subpoenaed inside the court's jurisdictional boundaries. If you reside in Houston, Texas, you cannot be subpoenaed to testify in a case in Tampa, Florida. As an Expert, your appearance is normally voluntary, and subpoena power is not an issue.

United States currency is attached to the subpoena—usually from one to fifty dollars. You keep the money; it is your fee for providing your services, unless (hopefully) you have negotiated with the parties for a more appropriate fee. Usually the party requesting the deposition pays your fee for expert testimony. That fee is limited to the time of deposition plus additional work requested during the deposition. Preparation time is not included.

Evidence

Some of the information you study during the discovery process has potential to be introduced as evidence at trial. You need to be aware of what constitutes evidence so that you do not overlook relevant documents and exhibits.

The jury's duty is to decide a case based on the facts. Since the facts are what is at issue, the evidence becomes the basis of the jury verdict. Documentation that is not introduced or accepted by the court as evidence cannot be used by the jury in their deliberations.

Evidence consists of oral testimony and tangible materials such as documents, exhibits, and demonstrative aides. A hybrid form of evidence is a deposition transcript. Admission of evidence into the court proceedings involves a two-step process. First, the evidence must be tendered to the court by the attorney. Then the court must act on its admissibility.

The burden of proof on the relevance of the proposed evidence is on the proponent. For example, the attorney introducing an Expert Witness must show that the witness is qualified. At the appropriate time the opposing attorney may object on some basis to the proffered evidence. The judge then rules, based on the relevance and materiality of the item. Does it speak directly to the issues in dispute? Does it help prove one of the propositions put forth? Does it make a supposed fact seem somewhat more likely?

If the evidence is admitted, the jury then judges the weight and believability of it. Thus, the ultimate determination of the relevance of the evidence is made by the jury in its deliberations. The judge is a gatekeeper, excluding evidence he or she believes shows unfair prejudice, confuses the issue, misleads or otherwise wastes time.

Also, the judge will exclude hearsay evidence. Testimony from a fact witness must be based on direct knowledge. Opinion is generally not admissible. The most important exception to the hearsay rule is the testimony of the Expert Witness. The Expert Witness can testify to opinions communicated by others and on evidence introduced during the trial. Frequently, tangible evidence is introduced into the court proceedings through the Expert Witness.

Commonly, the attorneys for both sides will have a meeting before the trial and decide what evidence will not be objected to during the proceedings. This allows the courtroom to function more smoothly. Items not agreed to will be objected to and fought over. Evidentiary arguments and the outcomes can be critical to a case.

As the consultant and Expert on the case you will be the most knowledgeable about the totality of evidence available. Your judgments about relevance of evidence is one of your most important duties.

Witnesses

The witnesses on your side are valuable sources of information and expertise. Meet with them separately and together. Find out their titles and understand their personalities. Have them recall the events, and how they really happened. Keep good notes. At trial you may have to state with whom you spoke and which information you accepted in developing your opinion.

You cannot talk to people working for the other side. However, you can visit with third parties, such as fabricators, service companies, construction firms, and medical personnel. You can use what they tell you in the development of your case. Often companies will not testify against former clients but will talk to you "off the record." You should never conduct any independent investigation or interviews, however, unless so directed by the attorney who has retained you.

Tell your lawyer who you think has the most information. He or she may wish to take that deposition first, before the other side has congealed their theories.

Talk to other well-known Experts in the field. Help your lawyer identify them. Your side can hire Experts first, before the other side. You never want the person who wrote the book to testify against you.

Other witnesses include independent witnesses and adverse witnesses. Independent witnesses do not work for any of the parties involved. They have greater credibility in the courtroom. Adverse witnesses work for the other side. They are called when they either are friendly or have essential testimony. Special rules apply in the courtroom regarding adverse witnesses. For example, the attorney may use leading questions to examine the adverse witness.

THE COURTROOM

Venue

Trial may be held in Federal or State Court. The location of the trial may depend on what the dispute is about: where the contract was signed, or where the negligence occurred. It is almost always to the advantage of one side to have the trial in their hometown.

Judge

Know your judge. He or she will rule on how broad or narrow your testimony can be. You need to understand the judge's reasoning. If they do not understand your points, they will tend to limit your testimony. What is their temperament? Relaxed or formal? Do they like jokes?

Judges tend to have two different approaches to evidence. One judge will allow everything, whether or not it is directly material, on the theory that this will be fair to both sides. That strategy limits potential appeals. Another judge will restrict what is admissible, and narrow the focus to move the trial along quickly. You need to know your judge's approach and calibrate your testimony accordingly.

Think of the judge as part of the jury. Your testimony will

have an impact on him or her. He or she may at times ask questions. At the end of the trial the judge will prepare the all-important jury instructions based on evidence and opinions presented at trial. The jury will decide the case based on the fabric of those instructions.

Jury Selection

In some state courts, the attorney can question each juror during *voir dire* as to where they work, where they live, their family, religion, union membership, etc., to elicit information about the potential bias of each person. In Federal Court, the *voir dire* is more tightly regulated: the judge asks questions and selects the jury. The attorneys can only suggest questions for the judge to ask.

The attorney can challenge a prospective juror two ways—a challenge for cause, and a peremptory challenge. The challenges for cause are unlimited. The attorney must show potential bias on the part of the juror. Each party has a designated number of peremptory challenges which have no requirements. The attorney may use these—based on intuition or whatever—to knock off potential jurors they feel would not be sympathetic to their side. The number of peremptory challenges can be expanded if there are more than two parties. Both sides have the same number of challenges. For example, if there is one Plaintiff and two Defendants, usually each Defendant gets three peremptory challenges and the Plaintiff gets six.

Plaintiff and Defendant

The Plaintiff has the burden of proof in the lawsuit. They must show with evidence how they were wronged. The Civil standard for proof is "preponderance of evidence," which in rough terms means that greater than 50 percent of the evidence is in his favor. The criminal standard is "beyond a reasonable doubt," which is gauged at greater than 90 percent proof for the Plaintiff to win.

Counsel and Cast List

You will usually be working with one lawyer. At trial, often a high-level member of the law firm will participate. They need to be brought up to speed late in the development of the case.

The other side's attorney needs to be scrutinized. Will he or she admit when they are wrong? Will they repeat a question exactly the same way the second time? Do they overintellectualize? Do they understand technical issues?

Witness Lineup

The lineup of witnesses testifying on your side is important. Your expertise is especially valuable here. The types of questions to ask are: What testimony is needed to fit into the legal and technical theories? How should the evidence flow? For example, you could start at the time the breach of contract or negligence occurred and work backward; or, start at the beginning of the contract and go forward.

What kind of case are you building—a technical one or a people one? Should the weak witnesses be put on the stand first, in the middle, or at the end? Usually, weak witnesses are placed between strong ones. Under the principle of regency and primacy, the best witnesses are placed at the beginning and at the end. Usually, the best fact witness leads off and the Expert Witness is the cleanup hitter.

THE TRIAL

Opening Statements

The trial begins with each side making an opening statement to the judge and jury. The attorneys stake out their positions and describe the evidence and witnesses each intends to have in court to prove their version of the case.

The Plaintiff's attorney then calls their witnesses. The "Fact" witnesses can only testify to facts to which they have personal knowledge. The Expert Witness can speculate and testify to probabilities. The Expert can also testify to hearsay, if it is the kind of information that would normally be relied upon in their business. The Expert is entitled—and expected—to give opinions. The fact witness, generally, may not.

Objections

When the opposing lawyer raises objections when you are on the stand, say nothing. Listen to the basis of the objection and how your lawyer responds. A common objection is to the relevance of your testimony. Can you work the same information into your testimony another way? Ask your lawyer up front if there is anything in your testimony that will be objected to. If you cannot get the information in on direct examination, maybe you can bring it up on cross examination.

Rebuttal

If you are an Expert for the Plaintiff, you may have another shot at testifying in rebuttal. Plaintiff gets to rebut the new points brought out by the Defense witnesses. What is allowed during rebuttal is within the discretion of the trial Judge, so you need to understand his or her attitude. Consider in advance the points which might be brought up on rebuttal.

Closing Statement and Jury Instructions

The closing statements are the lawyers' last messages to the jury. Tell your lawyer what evidence you think was persuasive to the jury and what evidence was not. What are the weak points? Experts can give excellent feedback in preparation for the closing arguments.

Jury instructions are usually prepared well in advance by both sides and submitted to the Judge, who explains the applicable laws and lays out to the jury what specifically must be proved. To be successful, your pleadings in court must be consistent with the judge's instructions. The jury then deliberates while you return to your regular life.

Closure

The case may not be over when the jury renders a verdict. Appeals by the losing party are probable. You may be called upon to evaluate the appeals briefs filed by both sides for technical accuracy.

Settlement negotiations may be recommended. Both sides are weighing the costs of further litigation against the potential upside and downside risks. These talks may involve technical points that have bearing on your expertise.

The old saying, "it ain't over 'til it's over," applies here. It ain't over until both sides have agreed in writing that it is over. Sometime, somewhere, someday, one of your cases will be appealed, and years later be set for retrial. Be prepared. Retain your records until your attorney gives you the word. Even then, do not throw away everything. Retain your reports, depositions, and trial transcripts of your activity. In future cases, your attorney will want to read the history of your involvements in prior cases.

3

Developing Winning Strategies

REQUISITES OF AN EXPERT WITNESS

Definition

An Expert Witness is a person who, by reasons of education or special training, possesses knowledge of some particular subject area in greater depth than the public at large. It is his or her job to assist the trier of fact (the judge or the jury) in understanding complicated subjects not within the knowledge of the average person. Most people are not allowed to enter the courtroom and give opinions. The Expert Witness is by virtue of specialized expertise present to do just that—give opinions.

The federal and state rules of court define an Expert to be someone with "scientific, technical, or other specialized knowledge." Does that necessarily mean degreed? Consider the source for expert testimony on how to saddle a horse properly; on the length of a vara (an early-day unit of length frequently used to measure land in Texas); on how a pressure cooker operates.

The avenues to becoming an Expert are varied. A trucker was recognized as an Expert in air leaf suspension systems, and allowed to testify, when it was shown he had been in the trucking business nearly 30 years and had controlled five tractors with air leaf suspension systems. He testified to the simple use of safety chains with these brakes. The bottom line is that special expertise, directly related to the matter at hand, must be clearly demonstrated.

The use of an Expert comes into play long before a case goes to trial. How and when the expert is involved depends on, among other things, their area of expertise, the type of matter to be handled, the purpose for which they were hired, and whether they have been identified by the employing attorney as an Expert Witness or as a consultant.

Functions of the Expert Witness

The court wants to hear from the Expert Witness to help the judge and jury's understanding of the case. Your attorney would like to have your opinion determine the jury's verdict. Between these two objectives, the Expert has a wide latitude in which to operate.

In offering opinion, Expert testimony performs four separate functions:

1. *Establish the facts.* You must first study the documentation in the case and decide what data and information directly bear on the matter at hand. Discovery usually yields a vast amount of paper. Sift through it and make a preliminary classification of relevance.

2. *Interpret the facts.* Tie together cause and effect relationships with the data and the facts. This will become the technical basis for your case. Do not be fooled by correlations that seemingly link cause and effect but have no theoretical justification.

3. *Comment on the opposing Expert's facts and opinions.* Develop a good understanding of the opposing Experts. Find out their educational background and experience. Read their publications. Probe for strengths and weaknesses. Oftentimes trials become a battle of the Experts. Prepare for the

battle with as much intelligence as you can gather. Take the opposition's Expert Report apart piece by piece. That document represents the other side's best case. Your attorney needs to know the most intimate details about the facts and opinions contained in that Report.

4. *Define the professional standards in the particular area of your opponent's expertise.* One of the most critically important ways an Expert is used in trial is to define the "standard of care" exercised by professionals in the field. Traditionally, standard of care has been defined on the basis of judgment normally exercised by professionals in good standing. Additionally, the professional must be informed or aware of the current practices. Obsolete practices are now considered by the court to be negligent practices. Thus, a professional must exercise reasonable, informed judgment in the course of his or her duties. You, as the Expert, will be called upon to define the professional standard and to measure that against the standard of care exercised by the professional(s) on the other side.

You can only testify in the area in which you are qualified. A chemist cannot testify as to engineering standards. However, a chemist can offer an opinion as to what happened but cannot discuss whether an engineer or other nonchemist negligently performed his or her duty.

Additionally, the Expert may be used to introduce fact evidence into the legal proceedings.

Hypothetical Case

Consider the case surrounding the death of a highly successful businessman killed in an automobile accident. Assume that the businessman was married with three young children. The surviving wife suspects the automobile may have been defective in some manner. She consults an attorney.

Several Experts may be employed in this case:

- Before a suit is filed, the attorney may consult with an automotive design specialist to determine the possibility of a design or manufacturing defect.

- Assuming there appear to be reasonable grounds for a suit, a psychiatrist or psychologist may be called upon to determine the degree of pain, suffering, and mental anguish of the family caused by the loss.

- Damages must be proven in terms of dollars; an economist calculates the financial loss, including lost wages for the years remaining in the husband's anticipated working life, the loss of service in household and other duties, and a variety of other services commonly expected of a husband and father.

- If manufacturing defects are alleged, an accident reconstructionist may be employed to reconstruct the accident to prove that, except for the substandard manufacture of the automobile, the accident would have been survivable.

- A Standard of Care Expert from industry may be employed to show that methods of manufacture of the automobile were below that of accepted practice in the industry.

- Clearly, it must be proven that our businessman died of injuries received rather than his having suffered a heart attack that subsequently caused the accident. Thus, a medical examiner or other qualified person confirms that the injuries received as a result of the accident caused the death.

Each of these and any number of other Experts may be employed at various stages in a case development.

Qualifications

To qualify as an Expert, you must have demonstrated mastery of a subject. Further, you must be able to defend your qualifications in court against the sharp questions of the opposing lawyer, whose purpose is to question or deflate your credibility in the minds of the jury.

It is important for you to gather your credentials into a well-documented presentation. Called a résumé or curriculum vitae, this presentation should contain all the facts pertinent to your background:

- Academic credentials—advanced degrees, special areas of study, patents, inventions.

- Experience—measured both in years as well as participation in noteworthy projects, patents, innovations.

- Associations—committee participation, chairmanship.

- Publications—have you authored papers in your area of expertise?

You do not need to cover all these areas. A truck driver with 30 years of experience can expertly describe the maneuverability of an eighteen-wheeler. Whatever your expertise, you must show depth and competence at a minimum.

The Bad Expert

Bad Experts have the following disabilities:

1. *Inexperience that shows.* They are nervous and tense around lawyers and unable to express themselves clearly.

2. *Talkative.* They do not know when to keep quiet. They answer questions that the opposing attorney does not ask. And they give too much information in response.

3. *Prejudiced.* They want to get even with the opposing Expert, lawyer, or litigant. They have ethnic or gender hang-ups that easily surface.

4. *Credentials are weak.* Their expertise does not come into direct alignment with the technical issues and they are too stubborn or embarrassed to admit it.

The Good Expert

The good Expert needs to be street smart and possess a strong theoretical background. Street smartness is just plain essential, since lawsuits generally arise from practical difficulties. The credentials of a good Expert should at least match those of the opposing Expert. In the battle of Experts, you do not want to be outgunned by a résumé that looks like the Manhattan phone directory. A good Expert acts and looks dignified, dresses in good taste, and has a confident demeanor and attitude. A good Expert is exactly that—*the* Expert in the field.

Courtroom Demeanor

The Expert must be able to tell two stories simultaneously. The technical story must hold up under attacks from the opposition as well as provide a record under appeal. You must also tell the story in terms the lay person can understand. It must be interesting and free of jargon.

If you lose the jury's attention, it is difficult to get it back, and you will also lose big points for your side. Look at the jury when you speak. Be animated. Use charts and graphs to balance your presentation.

Communication is everything. Do whatever it takes to develop a good communication style. Practice in front of a mirror and into a tape recorder. Have a nontechnical friend listen and critique your presentation. Keep the jury interested with your voice. Have them sitting on the edge of their seats, waiting breathlessly for you, the Expert, to explain just what happened and why. Be yourself—your most interesting and entertaining self.

Do not be proud of your knowledge or background. Just be matter-of-fact about it. Present yourself as an average person, much like the jurors, with some special expertise that will assist the jury in unraveling the mysteries of the case.

When you are on the witness stand your presentation needs to be totally jury oriented. The first five minutes of direct examination are critical in establishing your relationship with them. Make it a friendly time by establishing eye contact and talking directly to them. Sit up and smile. You are the Expert, the cleanup batter, there to provide the essential service of communicating a heretofore confusing matter in terms the jury can understand.

On cross examination, be deferential to the attorney whose job it is to destroy you. Again, you are the Expert, and he is the hired gun. *Never* argue with him. You can only lose, and your credibility with the jury will suffer. They begin to think you are not the unbiased Expert they were led to believe in.

You do have the right to fully answer any question asked by the opposing attorney. They may try to cut you off after your "Yes or No" response. Break in and indicate you have more to say. Explain your answers. If necessary, with the court's permission, step down from the witness box to point out an exhibit or chart that reinforces your point.

Deposition Demeanor

Deposition requires the Expert to assume a demeanor radically different from that required at trial. At the deposition, answer the questions in a truthful, direct manner with no commentary or hint as to the direction the attorney should go next.

The deposition is a discovery device. The opposition wants to find out your opinions and the bias in those opinions. Moreover, the opposing attorney will use the deposition to accelerate their learning curve on the subject matter.

You have no duty or obligation to provide educational services. Your objective is to respond directly and precisely.

You may frequently ask for a question to be restated or rephrased, if the attorney is not asking precise questions. Ask for more background. You should question the lawyer as to the assumptions you are to make in responding if that is not clear.

The deposition is attended only by the attorney and parties and a court stenographer. Objections will be made by your attorney at times. Generally, these objections are for the record and you will still have to respond to the question. A judge is not present to handle the objection. Respond after listening closely to the nature of the objection. Your lawyer may be trying to protect you. Follow your lawyer's instructions. If they tell you not to answer, do not answer.

Be sure to read your deposition carefully afterward. Sometimes a critical "not" will be left out. Remember, your reputation as an Expert is on the line in the transcript as a written, enduring record of your beliefs. Do not waive your right to review and sign the written transcript of the deposition. That is your protection against inaccurate or false representation of your testimony.

Interacting with Your Lawyer

Rapport with your lawyer is vital and essential. You need to be responsible for the scope of the technical case. Your lawyer must respect your opinions.

Your first duties are the exploration of the technical issues and the education of your lawyer. Attorneys like to play detective. But they will likely have already composed their theory on what happened and why, and may be just looking for you to tell that story. But too often the story is wrong. The client told the lawyer their story and it sounded plausible. However, most lawyers have minimal scientific training and want to believe their client's story. Here the Expert can provide a special benefit to the case, but be diplomatic. Be gentle. Be a teacher. Use evidence you gather to explain the technical theories. Break down complex phenomena into everyday metaphors. Practice storytelling on your attorney.

Cases usually start out looking quite simple. Both sides will quickly develop a logical explanation for what happened. You, as the Expert, will pierce the opposition's neatly packaged story of cause and effect and quite literally mess it up. Consequently, your lawyer will at first become insecure about the technical issues. It is only through education and confidence in you, the Expert, that your lawyer will learn to live with these insecurities.

Communicate frequently with your lawyer at first to develop the necessary rapport. This may be a challenge; lawyers do not want to spend much time on a case at the beginning. They have other cases in progress. Many times you will have to do the pushing.

The Attorney-Expert relationship may take long and arduous effort. But it is necessary if you are going to use your abilities more fully in the case. Establish your relationship on both a professional and friendship level. Lawyers are interesting people—like Experts—once you have broken through the jargon barrier.

Recognize that the issues confronting you and your attorney are never black-and-white. Technical professionals tend to gravitate to one point of view and negate dissenting opinions. Lawyers live in the world of grey, where facts can be colored and shifted into various points of view.

Flexibility is the essential ingredient. Allow yourself to inhabit the mind of your attorney and see how they construe facts differently. Certainly the other side, including the other Expert, will be doing just that. Play with the alternative opinions you and the attorney can develop to sharpen your findings and give insight into the opposition's case. Figuring out the other side's case is almost as important as knowing your own case.

PREPARATION OF TANGIBLE EVIDENCE

The Expert Report

Some months before the trial date, your attorney may formally designate you as his Expert by notifying the other side in writing. You will prepare an Expert Report of your findings and opinions. This document is very important. It exposes to the opposition the basis of your conclusions. They will be able to analyze the strengths and weaknesses of your position. It serves as a negotiating instrument for your attorney in settlement talks. And it also provides information to the opposition to aid them in preparing for your deposition and trial testimony.

As a result, preparation of the Expert Report requires much careful consideration. On one hand you want to elaborate the technical findings. On the other hand, the greater your elaboration, the higher the exposure to counterattack. Some compromise between minimal and full disclosure must be reached.

The flaws and contradictions in the technical aspects of the case are generally not dealt with here. Rather, the Expert Report represents your technical opinions offered in a consistent, forthright, affirmative, and truthful manner.

Brevity is essential. State your opinions, the reasoning behind them, and their technical bases. Write in tight sentences; the opposing attorney may (and probably will) try to lift certain passages from your report and construe their meaning in ways more favorable to their case. Inexact language allows for more opportunities for them to create other, unintended meanings.

Explain enough so that your opinions make sense. Flat statements without backup will cause trouble when you are testifying. Cover your tracks with a sound, logical train of follow-up sentences, data, theory, and references.

Avoid generalizations. Example: **"Experts in the field agree that _____ is true."** Have you polled the Experts? Do you have written evidence of other Experts' concurrence? If you have discussed the case with other Experts, you can say so. But be specific.

Reference your work carefully. You can inject statements from the evidence gathered (memos, correspondence, reports) and from recognized literature (journals, books, technical literature). Show clearly how your opinions mesh with the realities of the factual situation.

Add graphs and charts to clarify points. Underline or set out your opinions and conclusions so that readers can quickly grasp what you are saying. Ruthlessly eliminate typos, misspellings, and errors from the text. A sloppy report indicates a sloppy mind to the opposition. You can be sure the jury will be shown your imperfections.

Above all, make sure your attorney understands every sentence and every point. They will read the report from a lawyer's perspective. Expect criticism and be prepared to negotiate. The Expert Report must be able to withstand legal as well as technical attacks. The bottom line is that the report is *your* work. You must be able to stand behind every statement and every finding. Do not let your attorney stretch you beyond your limits, on paper or in court. You draw the line. Your reputation is at stake.

In Chapter 7, "The Engineer's Nightmare," two Expert Reports are shown. The first report, by Morris, is written in a conventional format. The first paragraph of the "Summary" defines the problem. The following paragraph lists the reasons for the problem with brief explanations. Dr. Morris then summarizes his opinion in the last paragraph. The next section is a narrative in which the case is explained from a historical point of view, using science and technical analysis. This report is the technical advocacy document Dr. Morris will be expected to defend at his deposition. It contains no extraneous information. The arguments are presented in a tight, logical sequence. Ambiguity is held to a minimum. Technical professionals reading the report will understand where Dr. Morris stands and why.

On the other hand, Dr. Lowry's Expert Report is vague and evasive. Most likely, Lowry's attorney has instructed him to divulge as little as possible in order to keep the other side in the dark. No evidence is provided to substantiate the opinions. You can be sure the opposing attorney will bore into Dr. Lowry about this at deposition or trial.

Jury Exhibits

Presentation of your findings and opinions is enhanced with charts, graphs, pictures, scale models, and other tangible items. These are demonstrative materials that do not have to be entered into evidence before you present them to the jury. (Any actual documents must be introduced into evidence before they can be used.)

Paint a picture for the jury. Before showing something, explain it first so that they will appreciate the information. Consider using colors in your exhibits. Red, green, and blue are the best eye catchers. Use overlapping transparent flip charts to show changes on drawings. Make sure the exhibits are enlarged so everyone in the jury can see and read along with you.

Slides are very good, and photos are especially valuable for showing the actual site, equipment, etc. Make sure the jury understands the visual perspective.

Reproduce and blow up the most telling evidence, such as memos, charts, and invoices that illustrate your points. Double check your exhibits for accuracy, your charts for typos, etc. Mistakes can cast doubt on your credibility and the impact of your chart could backfire.

Plan ahead of time what you want. Two to three months is adequate. The exhibits need to be organized with your lawyer. Before the trial, the court may require an exchange of exhibits between sides.

The value of exhibits to the jury cannot be overstated. Our culture lives in a visual world dominated by television, picture magazines, and other eye-catching media, whereas lawyers and technical Experts are surrounded by words and figures. You need to orient your thinking to a more visual mentality. Quality pictures are carried in jurors' minds long after transient verbal statements are heard. Imprint your story with high-impact representations.

Experiments

In a big case you may need to do experiments as part of the investigation. These experiments must be conducted with all the scientific precision of research. Furthermore, to testify about the results at trial, you need to be physically present during the running of the experiment.

THE ETHICS OF EXPERT WITNESSING

The use of the term "hired gun," as some Experts are called, is pejorative, implying that they are not objective but shade their opinions to mesh with the side of the case that is paying for their services.

In reality, the Expert is not a third party with total access to all information, making a total disclosure of all judgments and opinions. But neither are competent Experts shady characters who can shift their opinions to mesh with either side of an issue.

The Expert, by virtue of being retained by one side of the litigation, gets much more exposure to that side of the issue and is precluded from talking to witnesses for the other side and having discussions with the opposing lawyers. There is an imbalance in the quality and quantity of information received by the Expert.

Experts must be totally candid with their lawyer on the development of their opinions. The attorney needs to develop strategies and tactics to deal with the weaknesses of the case that the Expert exposes. For example, if the Expert finds the case to have many technical flaws, the lawyer may attempt settlement rather than hazarding a trial.

The attorney is the mouthpiece for the client. The Expert provides an opinion that the attorney will use to maximum advantage. It is in this use that ethical dilemmas arise.

Should an Expert expose weaknesses in the courtroom as a measure of his or her objectivity? To what degree should an Expert be led by the attorney? Certainly, Experts should never cross the ethical boundary of stating opinions that are not really theirs or that they do not believe in. But should an Expert withhold opinions or findings that are detrimental to his or her side?

Usually, this dilemma can be resolved on direct examination

by the attorney asking questions about the weaknesses. The Expert will usually explain these, using the words that are most favorable to the client and generally explaining away the weaknesses as "no big deal." How ethical is that? Are the weaknesses really no big deal? It becomes an issue of shades of grey, with the Expert again being pressed to shade favorably for the client. The Expert must draw an ethical boundary line. The worst case is when the weaknesses are so large that they are potentially devastating to the client. In that instance, the Expert has to decide whether it is in their best interest to testify at all. If the weakness is not brought up on direct examination, in all likelihood it will be brought up in the cross examination. And the Expert must answer the questions in a truthful manner.

How large a role should an Expert play on the litigation team? Most Experts feel "used" by attorneys, like tools in a toolbox. The tool is dragged out at the proper time and inserted in the trial to twist the bolt off, then discarded. This usage discourages beneficial interactions between the attorney and the Expert and heightens suspicions that the Expert is indeed a hired gun.

A closer relationship has many advantages. The Expert has the greatest grasp of the technical issues and can help the attorney develop a consistent story about the case. However, ethical paradoxes crop up. Should an Expert actively work for the team as an assistant coach or manager? Is this an appropriate role? Should Experts care if the side paying their consulting fee wins or loses? Or is the Expert there strictly to provide opinions?

There are Experts that actually promote cases and litigation. They clearly have problems with their objectivity when the courtroom confrontation comes. An Expert should be available to interact with the team and provide input all along the way. They are not divorced from the reality of the adversarial process. How much help they can ethically provide must be individually determined.

The ethical codes provide little guidance on the subject. For example, the National Society of Professional Engineers' "Code of Ethics for Engineers" states:

3. Engineers shall issue public statements only in an objective and truthful manner.
   a. Engineers shall be objective and truthful in professional reports, statements or testimony.

The degree of assistance to a litigation team is up to the individual.

SUMMARY

Keep in perspective that the lawyer is the Team Captain. You are an essential part of the team. You bring harmony and consistency to the case. Take the lead in the technical investigation; otherwise your lawyer will narrow your scope and limit your input. There is no worse feeling than one of being used as a tool in a trial. You can prevent that by integrating yourself in the team early on.

Do not be reluctant to bring out the detrimental side of the case. The warts need to be exposed so your lawyer can appraise the potential benefit for settlement and trial.

The legal process is a very human one, and it has its ups and downs. Your input and work are essential to the process.

In Appendix B is a copy of the "Recommended Practices for Design Professionals Engaged as Experts in the Resolution of Construction Industry Disputes." These guidelines are recommended by the National Society of Professional Engineers to help the Expert maintain an unbiased approach in dispute resolutions.

4

Marketing Yourself

Almost everyone has interfaced with the legal system at one time as a juror, in court fighting a traffic ticket, or as a party to a lawsuit. The experience may have left you with mixed emotions. You were involved with the legal system without a broad understanding of it.

Now you are considering marketing your services to the foreboding and somewhat alien legal profession. They and the clients they represent will be your targets.

You need to have a marketing strategy that represents your expertise in the highest professional way. Lawyers want the best Experts they can afford, and "best" includes your professionalism. The way you market yourself is the initial impression your potential clients will have of you.

Generally, the attorney and client will develop a list of Experts jointly. The attorney will make the final decision with the consent of the client. Your marketing strategy must be two-pronged, aimed both at attorneys and at your own profession. Leads come from both directions.

The likelihood is that you have established a reputation as an Expert in your profession. But is it known that you want to be an Expert Witness? In your normal dealings with colleagues you can indicate your availability if the opportunity arises. Ask your current clients about pending or ongoing litigation and what attorneys they use, and if they know of lawsuits in your area of expertise. Let the word out, and reinforce your interest without being obnoxious.

Using your existing network is the easy side of marketing. Getting to the lawyers is much tougher. Much of this chapter is devoted to that side of the client list. The first issue is how much to charge. If you charge too little, no one will think you are an Expert; if you charge too much, no one will be able to afford to use you.

FEES

Schedule

Your work in preparing an estimate of the cost of your services will be greatly simplified if you have established a schedule of fees for various services. Such a schedule might include, among other items:

1. personnel (rate per hour)
   - principals
   - others

2. service categories
   - court appearances (full- or half-day increments)
   - field investigation (per hour or day)

3. miscellaneous services
   - services of other consultants or subcontractors
   - rental or construction of special equipment
   - construction of models or exhibits
   - storage of evidence

4. miscellaneous charges
   - transportation (airfare, mileage, parking, etc.)
   - lodging and meals
   - telephone charges

Most consultants charge an hourly or per diem fee that they apply to all of their work. Sometimes, a range of fees is used for various services, with service categories and miscellaneous charges billed at cost plus a percentage to cover administrative expenses. Others charge the same hourly or daily fee regardless of the type of service. Some charge time and a half for work on Sundays and holidays or for work in excess of eight hours per day. Few consultants charge on a lump-sum or fixed-fee basis because of the difficulty in estimating the amount of time and work that a given case will involve. You should choose the fee basis that best suits you and your circumstances.

Retainer

Like attorneys, most forensic consultants require the advance payment of a retainer fee, billing against it until it is depleted, at which time periodic billing begins. If you have definite requirements on these issues, make them clear at the onset.

Insist on having a signed contract, letter of agreement, or other written confirmation of the engagement of your services before beginning work on a case. There are a number of ways in which this can be accomplished. The attorney who has retained you may draw up a contract. Read it carefully to be certain that it includes all the terms that have been agreed to. Be sure it includes a specific statement that payment to you is not contingent upon the outcome of the case. Alternatively, you may prefer to use a standard contract with all clients. You may wish to consult an attorney on the form most suitable to your circumstances.

As a general rule, all of your services will be charged to your client, with the exception of your deposition, which is paid by the opposing side. Although in some cases your client will be responsible for your deposition fee, usually the attorney who requests the deposition and asks the first question is responsible for compensating you. The fee you charge will be for the actual time spent giving the deposition plus travel time. The fee you charge must be your normal fee, whether hourly, or in half-day or full-day increments. If the reasonableness of your fee is challenged, the court may rule on the amount you are allowed to charge. The time spent in preparation for a deposition is charged to your client rather than to the opposing attorney.

You may submit a statement for your time to the attorney who requested the deposition. Alternatively, you may submit an estimate based on the anticipated length of the deposition and require payment in advance. If the deposition takes more time than originally estimated, you may bill later for the difference between the advance payment and the total fee. The attorney who has retained you can advise you on the procedure for advance billing for depositions.

Fee Structure

Your fees as a consultant should be in the same range as your normal charges for services. The charges for services as an Expert under oath can be at a premium, ranging from 50 percent to 100 percent over the base fee. In actuality, the time under oath is very small compared to the total time spent. However, waiting time and time spent in preparation for testimony can also be charged at the premium rate.

One simple system is to charge a flat hourly rate at some increment higher than your usual rate. "Time spent" is that period taken away from other matters for the case. Travel and think time are included in that category. Expenses can be billed at cost plus some small service fee, usually 10 percent.

Billing on a monthly basis is a standard procedure, and an important habit for you. As a rule, if you are billing time, you need to keep your lawyer informed so the bill does not come as a surprise.

POTENTIAL CLIENTS

Criteria

Investigate criteria for listing yourself in various local and national organizations, both general and specific to your area of expertise. There are several, including The American Bar Association's Register of Expert Witnesses. The Defense Research Institute, a Chicago-based organization of attorneys, has established an Expert Witness Index. The Forensic Services Directory is maintained by the National Forensic Center, which also supplies lists of Experts to Westlaw's computer-based Forensic Services Directory.

Professional Societies

Some professional organizations maintain lists of members available and specially qualified to serve as Experts. For example, a listing in *Engineers of Distinction: A Who's Who in Engineering* may prove helpful for engineers. Seek and maintain membership in professional societies; investigate the methods they use to handle inquiries from the general public regarding availability of qualified Experts. Make sure those who respond to such inquiries have at hand your up-to-date vita.

Expert Witness Service Companies

There are organizations whose function is to match attorneys and other prospective clients with appropriate consultants. These organizations will usually be listed in the yellow pages under the category "Attorney Service Bureaus" or something similar. While use of these agencies represents increased cost to the attorney and their client, the expense may be justified by the efficiency of the search. The Expert benefits from such a service in that the broker finds the referral and manages the administrative details. The services will add a 10 to 20 percent fee on top of your rate, and this is paid by the attorney. When working with these organizations, however, you should evaluate each case independently, declining those that are inappropriate to your expertise.

Networking

Get involved. Become politically active. Attend city council meetings concerning matters of interest to you or which involve some aspect of your specialty. Be prepared to interject your point of view. Become a familiar face in the political arena—where there's politics, you'll find attorneys. Such participation can provide exposure to potential consumers of your services while you shape your own environment through politics.

Utilize your current networking techniques to spread the word that you are interested in providing expert testimony. Word-of-mouth advertising is a very effective marketing tool.

Letters to Attorneys

An important target of your efforts must be attorneys themselves. Do you have friends who are attorneys? Who handled the incorporation of your business? Who prepared your will? Who writes your contracts? While attorneys who provide such services may have no need for expert testimony, they will surely have friends and associates who do, and can put you in contact with them with an introduction.

Advertising/Direct Mail

Advertising is more direct, but expensive. Review the content and appearance of ads in current issues of state and local bar journals. The placement, wording, and overall appearance of your advertisement are all important. If your budget, level of training, and expertise justify the expense, consultation with an advertising firm that specializes in the legal area is advisable.

Also direct, but less expensive, is a mail campaign aimed at the larger law firms, which commonly maintain a catalog of known Expert Witnesses. One way to begin is to send out a brief letter (snappy but informative) with a current résumé or vita and perhaps a few business cards. Although you may receive only three or four inquiries from a mailout of 500, that may be enough to justify your investment.

For a mailout, names and address can be collected from a variety of sources. The Yellow Pages telephone book in larger metropolitan areas will categorize law firms by the type of law practiced. State and local bar associations print directories and yearbooks that are a useful resource. For example, the state bar association publishes a legal directory that is on file even in small firms and may be found in the local law library. Most state and local associations print a yearbook. Some state associations have available for sale a listing of licensed attorneys organized by categories.

If you wish to solicit business in another state, contact that state's Bar Association regarding availability of address lists. Another good source is the *Martindale-Hubbell Law Directory*. Medium-sized and larger firms are usually subscribers to this yearly publication, which you might obtain through an attorney friend, and which is also usually kept in the local public library.

A directory of organizations is in Appendix A.

SUMMARY

Marketing yourself in the highest professional manner is essential. The marketing effort must be systematic and regular. Lawyers hire experts when the need arises. Every three to six months the legal community needs to know about your services. Make sure they know who you are and what you can do. Persistence, one of the attributes that made you an Expert, will make you a sought-after Expert Witness.

5

Discovery

DOCUMENT PRODUCTION AND ORGANIZATION

Most cases are won or lost during discovery. Sometimes key information is never properly requested from the other side; key documents may not be identified or, conversely, there may be so many documents you cannot keep track of them; depositions, also a part of discovery, can be minor skirmishes or major battlefields, depending on how the discovery material is utilized. Depositions are very much impacted by the rest of the discovery process. To be successful, you must know the strategy and tactics required to organize yourself and your information most effectively. In this chapter you will be taken step by step through the discovery process, learning the details of document production involving negligence issues.

As mentioned in Chapter 2, the initial stage of a lawsuit is discovery, following the initial complaint and answer. During discovery each side is requested — and required — to give the other side access to information, documents, and key personnel with pertinent facts. Thus you will be given documents from the files of both parties to examine.

The discovery process often begins with one side asking the other a set of written questions, called interrogatories, briefly discussed in Chapter 1. You may begin your participation in the case by assisting your attorney prepare the interrogatory questions — or the answers, depending on whether you have been hired by the plaintiff or the defendant. Interrogatories must be answered in writing by a certain date and are commonly used to establish basic factual information such as the names of personnel in specific job titles for specific periods of time; location of manufacturing facilities; manufacturing processes; raw materials used, etc. Because the answering party will only give the information asked for, it is important the questions be carefully crafted; and the assistance of an expert can be invaluable.

At about the same time early in a case, the plaintiff will file their initial request for production identifying types of information to be provided/revealed by the defendant. Experts can also be needed to properly identify the information being requested, since it is within the documents of the defendant that the plaintiff must find evidence of negligence. The defendant also files interrogatories and request for production. In a complex case there are a series of interrogatories and requests for production filed by one side or the other, as legal skirmishes are won or lost and additional information or issues are identified. The war of discovery is a paper war between the plaintiff and defendant, but it is certainly a war — which often determines the final outcome of the case.

As an expert you will be sent all or some of the discovery documents: memos, reports, consulting documents, doctors' reports, news clippings, etc. and perhaps the response to your side's interrogatories. In case you did not assist in the preparation of your side's request for production and interrogatories, it may be useful to request a copy to check that all the classes of information important to *your* investigation have been specifically requested.

Discovery Strategy

There are basically two strategies employed by the defendant's law firm: feast or famine. When the 'famine' strategy is employed, documents are only produced to the other side when they have been so specifically requested they cannot be withheld or the court compels their disclosure. Obtaining any *useful* documents is the challenge. Whether you are working for the defense or the plaintiff, the wording of the interrogatories and requests for production are very carefully written and/or read; the second and third round requests will hone in on particular areas of information. This strategy often employs entire classes of information — complete with elaborate indexes — which the defense may declare to be privileged documents. Only the judge can rule on the relevancy of privileged documents. Judges commonly allow cases to proceed with minimal court direction because they are very sensitive to issuing any directive which could be grounds for appeal later.

Conversely, when 'feast' is the defense strategy, the expert's role becomes the classic search for the needle in a haystack. Boxes and boxes of documents, most of the material of little or no interest, create a document bottleneck which can prevent the really valuable materials from being found.

Strategies can change as discovery proceeds, material is produced and the court rules on issues brought by one side or the other. These events can also dramatically affect the momentum of a case — a plodding-along investigation can become a whirlwind of effort, should the other side request a hearing on issues requiring your testimony, information you are developing becomes essential to another expert's research, etc. It is imperative that you, as the expert, are organized and responsive.

Bate Numbers

To keep track of documents, the law firm normally prepares an index of all discovery documents (sometimes the court requires each side to provide one). The index organizes documents by title, author, date, and, very important to the legal profession, a "Bate number", named after the fellow who invented the sequential-numbering ink stamp. This small seven-digit number, usually found on the lower right hand margin of each page, gives each sheet of paper a sequential number. There may also be a three letter prefix to the number which identifies the case and who produced the document, the defendant or the plaintiff. For example, in the hazardous waste case involving a suit against Environmental Perfumes, if the defendant invited the plaintiffs to visit their Chicago offices to review material, their documents would be identified with the letters EPC = Environmental Disposal/Chicago discovery. The three-letter prefix can be very valuable in identifying the source of a document.

The Bate numbering system can be especially helpful to experts. Not only can it be very valuable to identify the exact source of a document, but also any document, or any page of a document can be quickly identified and found just by knowing its Bate number, since the law firm usually keeps all these documents in Bate-numbered sequence. There is a downside to this system, however; this is normally the firm's *only* organization of discovery material, and related documents which enhance each others' meaning are frequently separated by hundreds or thousands of Bate numbers. Here is where the expert must know how to organize the material for his own evaluation and use.

Your Organization

The organization of material is critical to the expert's effectiveness. First, and most importantly, you must make sure you have the information you need upon which to base your opinion. Which one or more of the four elements of negligence discussed later in this chapter will be proven by your testimony? Each element should be a separate file, with an index of documents to readily identify and keep track of the strengths and weaknesses of each element on which you will render an opinion.

Almost as important as the search, your organization must allow you to retrieve and keep track of critical documents. Especially when you are being inundated with a river of materials, it is important to know how to fish out what you need. Just knowing that a lynch-pin document is in one of your boxes of papers is not enough; you must *find* it and tie it to other key documents, to build the story — *your* side's story — of what happened. The most effective way to build an index is to enter key information (title, date, source, area of interest, Bate number, for example) about the individual documents into your computer. Most word processors have some search capability which you can build into your index, since you know the documents and will know the document you are seeking. A data base like Lotus or Paradox can give you needed organizational breadth and depth, if you have a great many documents. Either way, I cannot over emphasize the importance of having a system that you can handle, and will use.

Develop documents that will help you communicate with your attorney, to assist him/her wed the technical side of the case to the legal side. These documents, illustrating one or more of the elements of negligence, can be a timeline chronology of events; the listing of toxic or hazardous chemicals involved and where they were found; the regulatory history of the plant; process flow diagrams, etc. It should be information that lays the technical foundation for your side of the case. This is the kind of information that your attorney needs when asking for material, and is especially useful to have when he/she is taking the deposition of the opposing expert. (That deposition is as important as your own, and if you are not be in attendance, you may also be asked to prepare questions for your attorney to ask.)

Finally, ancillary to your evaluation of discovery documents, it is important to ensure that your investigation fits an objective view of reality, by identifying all possible *other* sources of information about the case, discussed in detail in the next section: regulatory agencies, national standards, government funded research, etc. If you are being buried by documents, these outside sources can become a separate index of reference material.

PUTTING THE ORGANIZATIONAL TOOLS TO WORK

The expert creates order out of chaos by putting these organizational strategies to work. But first he/she must have a basic understanding of the legal framework, i.e., the legal issues that must be proven or disproven by the plaintiff or the defense. This critical guidance to the expert's investigation can be demonstrated by walking through the elements of a negligence case.

DISCOVERY IN NEGLIGENCE CASES

Negligence has a very specific connotation; one person (the defendant) is accused of allegedly harming another person (the plaintiff) because an accepted code of behavior — the *standard of care* — was not met, and which caused quantifiable damages. The burden of proof is on the plaintiff. In order to win the lawsuit, the plaintiff must prove with a preponderance (51 percent or greater) of evidence, the folowing chain of events:

1. The defendant had a duty, *and*
2. The defendant breached that duty, *and*
3. The breach of duty caused harm
4. Which resulted in damages

If the defendant can disprove at least one element in the proof, he wins.

Each link of the four elements needed to prove negligence is subject to review by technical experts. Upon being retained as an expert by the attorney, you will discuss the element(s) of the lawsuit in which you will be involved. In small lawsuits you may be asked to offer opinions on all four elements. In a large case involving complex causation issues, many more experts may be involved. For example, in a big hazardous waste site case, one expert will opine about the company's duties to dispose of wastes properly. Another expert will be called upon to form an opinion if or how wastes escaped from the site. A toxicologist will deal with the chemical impact on human health, and an economist will describe monetary damages. Commonly both the plaintiff and the defendants will employ identical "stables" of such experts to work the important areas of the case which relate to the four elements listed above.

THE CASE

The following describes what to look for, to prove or disprove the four essential links in a negligence case.

Duty

The first two elements, duty/breach of duty is proven (or, in the case of the defendant, disproven) by looking at four basic areas of responsibility that must be met:

Failure to Meet the State of the Art

This means the advice given by the defendant to the plaintiff was inadequate, obsolete, low quality, or just plain wrong. For example, was a water sampling procedure specified with too low a frequency of collection? Was a machine specified without adequate spare parts? Was a decade-old computer program recommended which was too slow for the specified computations?

Failure to Notify or Disclose Dangers

Was the advice given to the plaintiff unreliable or unsafe? Or did the defendant not disclose safety information he was aware of? For example, the air conditioning system cooling tower in an office building sprayed the auto parking lot with a salt mist. The designer perhaps knew this but failed to disclose it to the user.

Failure to Predict Adverse Consequences

The defendant may have been ignorant of negative potential problems. Should the defendant have reasonably foreseen bad things happening not only in ordinary use situations, but under conditions of improper maintenance and operation? In the operating manual, did the manufacturer of a gas turbine warn the owner about the consequences of overfilling the lubrication system?

Defining Duty and Standard of Care

The standard of care for non-professionals has long been established by court precedent to be "the behavior of ordinary reasonably prudent people." If one violates this standard in driving an auto, burning trash, or performing other normal living activities, he can be found negligent and be responsible for resulting damages. Under this definition, the ordinary person is not required to have any special training or education, but to uphold normally accepted standards of behavior.

For the professional, however, the standard of care is significantly different. The professional is supposed to possess special skills and knowledge to perform his job. Thus, the duty must be calibrated against the acceptable conduct as practiced in the profession, which the expert must judge and be prepared to offer an opinion if called upon. The issue becomes one of measurement of professionally acceptable conduct by the expert. The historical standard of "ordinary and reasonable skill exercised by one of the profession" was derived from the 1896 case of Coombs vs Beede in which the court of appeals stated:

> "unless he represents that he has a greater or less skill or knowledge, one who undertakes or renders services in the practice of a trade or profession is required to exercise the skill acknowledged normally possessed by members of that profession or trade in good standing."

With the rapid expansion of information and technology experienced in the recent times, Coombs vs Beede has undergone significant alteration. Back in the 1896 era and before, professional practice was one of stability with slow evolutionary change. Professionals were generally craftsmen trained in traditional skills. The 20th century knowledge explosion forced the law to cope with the impact of formal education and the rapid formation of new information and obsolescence of old information. Importance has to be placed on how informed the professional is expected to be. No longer can a professional rely on traditional knowledge. He has to stay abreast with the knowledge current in his field. He has to be "informed."

The question for the professional then becomes: What knowledge is he *expected* to possess? There are five areas where the professional can be "graded."

Cutting Edge

This knowledge reflects the latest research. It is untested in the real world and likely to be controversial. It may or may not be valid new knowledge. Only time will tell. Expect to find this knowledge at research symposiums. An example is the impact of low levels of ozone on human health.

Literature

This is knowledge that has been cross checked by a number of researchers. Some researchers find the knowledge to be valid while others are in disagreement. Literature knowledge is characterized by the publication of opposing viewpoints and contravenes. Usually this knowledge has not been field tested and is being subjected to questions of validity, limits, effectiveness, and economics. Expect to see this knowledge at professional meetings with sessions on new developments. The use of genetically altered microbes to degrade polychlorinated bi-phenols (PCBs) is an example.

Professionally Accepted

This knowledge has a widely accepted theoretical basis and is being used in full-scale applications. The literature is replete with articles explaining peculiar applications. The use of air stripping for the removal of volatile organics from a waste stream is an example.

Competency

This knowledge is learned at a college or university. It is found in textbooks and design manuals. This traditional knowledge represents the background acquired by professionals through the formal education process.

The professional must not only possess competency knowledge but must be conversant in professionally accepted knowledge. He has a duty to stay informed and incorporate current scientific and technical knowledge in his thinking, advice, and work. As previously discussed, scientific and technological knowledge is evolving so rapidly that the methods used by professionals quickly and sometimes radically change. His duty is to keep abreast of these developments. If he does not, he may be found guilty of negligence and malpractice. What was excusable ignorance yesterday is negligent ignorance today. Accordingly, the legal terminology now incorporates phrases such as "knowing, or should have known" as the standard.

As you go through the paperwork, and talk with your attorney and fact witnesses, look for the following:

Were the Professionals Qualified?

Did they possess the appropriate academic degrees? Were they working in the fields they were trained in? Did they have sufficient experience? Were they up to date in knowledge?

For example, an environmental engineer is not a biologist or geologist. However, opportunities present themselves to stretch the expertise of an environmental professional to an allied field. That is the way new things are learned. The question is, when did the professional exceed his educational limits and put himself into jeopardy? For example, a biologist was working on the fauna and flora of a potential hazardous waste site. He was asked to head up a study concerning bioremediation. Was that stretching too far? That question is not easily answered. What is the status of his knowledge and experience in bioremediation? Does he work on other studies involving bioremediation? Is the study science or engineering oriented? Have other biologists with similar back-grounds taken on this responsibility? Most importantly, does he meet the educational requirement to have a threshhold level of professionally accepted knowledge? Does he belong to the appropriate technical societies, attend the meetings, subscribe to the journals, go to the short courses?

Professionals always run risks taking on duties in new areas. These risks can be minimized through competent supervision. At a certain point of experience he can become a professional in an allied field. Chemists can learn engineering through experience, education and test taking. Engineers can develop into scientists over time. The degree of competency, education, and experience required of an individual to make a professional transition is what you as the expert must decide.

Did Colleagues Review his Work?

Peers and supervisors need to internally review work. Every-one makes mistakes; no one is infallible. A good internal review will uncover errors as well as question assumptions and conclusions. No document should be presented to the outside world without review. Internal review minimizes errors and spreads the risks and responsibility for the work in an appropriate way. Was the review process delineated in a best management practices manual? Was the manual normally followed? Were the reviewers knowledgable and competent?

Were the Advice or Recommendations Qualified?

Were strengths and weaknesses delineated in the documentation? Do they rely on documentation from others and state as such? Are the uncertainties and unknowns delineated in such a way that the interested parties had the opportunity to reject the advice or knowledgeably accept the risks? Advice concerning high risks must be extensively documented. Was this done?

Were Progress Reports Provided?

Were the interested parties informed about significant events? Many lawsuits evolve simply from interested parties being kept in the dark. Malpractice is often alleged because the work was not finished on time as promised. Was the client notified at every stage through written communication, explaining what happened and why? Were site visits, meetings, and phone conversations adequately documented, with copies sent to those who needed to know? Nothing is more powerful to a defendant than when an expert finds documentation supporting the thesis that the plaintiff agreed with important decisions which are being subjected to litigation.

Were Any Irregularities or Deviation from the Standard of Care Promptly Reported?

The natural urge for a professional is to attempt a fix-up of any unexpected problem or emergency that appears before anyone on the outside knows about it. However, such actions without notification look like cover-ups. Was there prompt oral transmission followed up with a memorandum? A jury will look favorably on a party that acted quickly and responsibly and notified relevant parties, particularly in writing.

Were Reasonable Alternatives Discussed?

As previously mentioned, one source of lawsuits is the situation in which the designer selected obsolete or inappropriate technology. Client participation and approval coupled with documentary knowledge of the options available blunts that argument. Did the professional take care in informing the decision makers of all the options available?

What was the Professional Competence of the Staff and Management?

Compare the job description and kinds of people in the job slots in the organization with the competitors. Did they meet industry standards of care in organizational management? Did they have a training program?

Did They Have a Set of Formal Rules of Conduct and Have Employees Abided By Them?

Bureaucracy does have an important role to play. For example, designs need to be formally checked, and drawings formally approved. The organization needs to have rules that define how it operates. Purchasing, billing, checking, paper flow, organizational chart, and decision making rules must be written down and followed.

In smaller organizations work is done more informally. Nonetheless, design documents should be signed and dated, and paper flow regulated. Smallness is no excuse for not establishing rules of business. The rules may be more informal and fewer in number. A truly professional organization looks, acts and is one in every respect. A jury will look for evidence of professional organizational behavior.

Were the rules followed? Are the rules reasonable and within the context of the ability of the people to follow without hamstringing performance? Most jury members will have worked for organizations and will be sensitive to rules and how they are interpreted.

Were State of the Art Facilities Maintained?

Were the laboratory equipment and methods obsolete? Computers play an important role in the life of organizations. Was the computer and software used by the organization state of the art? State of the art refers to the time of the negligent acts, not today.

Were Outside Consultants and Subcontractors Screened Adequately?

Many problems develop from the use of others to perform work. When something goes wrong involving a consultant or subcontractor, the main contractor is blamed. Were the contracts worded to define the responsibilities of the parties?

These guidelines all involve standards of care, which are at the heart of negligence lawsuits. You as the expert must be able to define your positions on professional duty and how each party to the lawsuit measures up. Your professional conduct and experience will be the basis of your decision-making process. Be aware your background will be scrutinized; and be prepared to explain how you fulfilled your responsibilities in practicing your profession and how you used your past as a benchmark for your opinions.

Causation

Causation is the third element in a negligence case. In the first part of the investigation, you learned how to figure out "who done it," now you need to find out *how* it was done — the technical detective work. The standard which you must use to base your opinion is, "beyond a reasonable degree of scientific certainty." What that means in a particular area of expertise depends on the area of science and the rigor of proof required. Some disciplines require higher degrees of confidence, others lower. Psychologists testifying as to the mental state of a person do not deal with statistical probabilities. However, an expert chemist may confidently opine on the purity of a polymer in statistical terms. The bottom line is: your opinion must be based on probabilities, not possibilities. Opinions based on possibilities are automatically thrown out by the court. How probable the causation must be is difficult and is usually left to the jury to decide.

The investigative tools you use to work on the causation issue will be subject to great scrutiny by the other side. Use the standard and accepted tools of your profession to minimize controversy.

The following is some helpful guidance in doing the causation analysis.

Be Open to All Possibilities

At first come up with several different scenarios. It is unlikely you or anyone involved with the case on either side will ever know absolutely what happened. Big holes will always remain, leaving you to speculate and develop your own assumptions. Test alternative scenarios as you go until you feel most comfortable with one or two. However, always remain open to surprises as witnesses are deposed and more information is gleaned. You can change your mind many times; and only at trial on the witness stand do you need to present your definitive theory of causation. It's nice if it matches the theory you gave at the deposition; but if facts become known after that, you can explain to the jury why you changed your mind.

Play Expert for the Other Side

Look for clues that help the other side and develop theories of causation they could use. One of your most important roles will be to shoot holes in the other side's case. You prepare for that during research. Furthermore, looking at the case from the opponent's eyes helps develop your theory. You will need to account for the outliers and the data that does not fit. One of the great battles experts wage is the war of theories. The more you know and understand about the opposing one, the better you will be on the witness stand.

Consult with Other Experts

Test your theories with other experts, and have the law firm pay them for their time. These conversations are a part of the collegial experience technical professionals normally participate in. They will help you clarify the strengths and weaknesses. And in deposition or at trial a little name dropping is advantageous. It shows further verification of your theories by other notables in the field.

Dig, Dig, Dig

Persistence is a prime characteristic of successful experts. Great detective work demands much gumshoe and burning of the eyes. A time tested way to make the most convincing technical argument is to know more about the case than the opposing expert. Greater knowledge shines brightly through to the jury.

Damages

The fourth element to be proven in a negligence case is damages, expressed in dollar terms. In larger cases an economics or accounting expert is called upon to define monetary damages. You as the technical expert will set the stage for this important testimony by delineating the specific areas of damages, or arguing that damages are not as the other side has said. You may be called upon to give damages estimates if you have professional competency; for example as a cost estimator.

Attorneys like experts with multiple skills. It cuts down on the cost of experts, and minimizes the potential for multiple experts to contradict each other. Just be sure you are comfortable with testifying about damage estimates.

Plaintiff attorneys love to have as high damage estimates as possible for negotiations. Do not get caught up in big numbers and make sure the figures are defensible.

CONCLUSION

The truly great fun of being an expert witness is playing the part of a technical Sherlock Holmes. You get to try on various roles as you attempt to put yourself in the shoes of the participants to understand what happened, rummage through files in search of smoking gun memos, interview (your side's) fact witnesses, read depositions, and visit the scene. And like Sherlock Holmes, you must be willing to get inside the case, and carry the facts of the case around with you in your head, like an invisible jigsaw puzzle. When my mind has been well-fed with information, I enjoy arranging and rearranging the uncertain areas in my mind during odd moments: standing in line, boring plane trips, elevator rides. I have found this method a good one to reveal new questions and answers about the case, which then can be subjected to the methodical investigation outlined above.

Discovery is when you must put forth maximum effort, using all the qualities that made you an expert in the first place. Like getting another advanced degree, the case requires you to develop information that will stand up to objective examination. Both your investigation and your conclusions will be tested again and again by the other side during deposition; and, if the case does not settle, you will be privileged to participate in one of America's great constitutional activities, the public trial.

6

Giving Testimony

Giving testimony is your most important function as an Expert Witness. You are ethically obligated to tell the truth at all times, which will reflect your high character and goodness as a human being. The opposing attorney will be searching for any deviations from the truth and for ambiguities in your testimony. That is their job. As long as you have performed your investigations under the highest professional standards, and have arrived at your opinions through careful and thoughtful deliberation, you have nothing to fear.

DEPOSITION TESTIMONY

Impeachment

From the opposing attorney's point of view, the purpose of the deposition is to create a record for future impeachment. The questions will be directed to four areas:

1. What are your credentials?

2. What did you do?

3. What are your opinions?

4. How did you arrive at these opinions?

It is not appropriate for the attorney to cross examine you. No useful purpose is served by impeaching you at deposition. If that happened, your attorney would simply find another Expert and you would never see the light of the courtroom.

The deposition environment can appear to be lighthearted and informal. It is usually held in the conference room of a law office, with the lawyers bantering back and forth. Do not be misled. Take the proceeding very seriously. Every word you utter will be recorded. Your effectiveness in the case depends on the outcome. Be attentive and be serious.

Some guidelines are: Keep in mind your client's theory of the case; and second, avoid educating the opposing lawyer, as discussed in Chapter 3. The opposing counsel will want to know what is favorable to their case and what is your theory of the case. They will, at times, ask outlandish questions. Resist the temptation to use your superior knowledge to set them straight. Remember, you are not there for any reason other than to answer questions. Rely on the attorney who engaged you to control the scope of the deposition.

The deposition process can be a lengthy one. Be patient. If you appear to want to bolt from the deposition, cut it short, or look uncomfortable, the opposing attorney will jump at this clue and try to drag out the deposition. The strategy of bearing down where you seem most uncomfortable is a psychological technique used to force inappropriate answers from you. Hide your desire to get it over with. Show toughness and resolve and a willingness to stay as long as necessary.

Other suggestions: Do not be sarcastic, flippant, or cute. Pause between questions. You are not less of an Expert if you pause. Do not allow the opposing attorney to develop a cadence or rhythm. Control the questioning process with your slow, deliberate, thoughtful answers.

The psychology of examination by the opposing attorney is to control the action. They will start off the deposition with ten to fifteen questions that are easily answered with "yes" or "no." They are trying to develop a friendly rapport with you. Answer the questions tersely with no facial or body language.

Typical Questions at a Deposition

The questions will start with a probe of your education, experience, and background. You should answer without elaboration.

Q: What universities did you attend?

A: Michigan and Iowa.

Q: What degrees did you receive?

A: B.S. and Ph.D.

Q: What were your majors?

A: Chemistry and Biophysics.

Q: When did you graduate?

A: B.S. 1962; Ph.D. 1970.

Further questions:

Q: Where have you testified before?

There are libraries of prior testimony; if you were previously involved in any significant cases, the likelihood is that the opposition will obtain access to your prior statements under oath.

Q: What did you do to prepare? How many hours did you spend in preparation?

Q: Did you undertake any research?

Q: What work did you do?

Q: What information did you review?

Q: Have you collaborated with anyone?

Q: Have you visited the site?

Q: Who did you talk to?

Q: What is your opinion of the opposing Expert's reputation?

Q: State every opinion you have formed.

Q: How much are you being paid?

The opposing attorney will go fishing in many directions and cover as much territory as possible. Your Expert Report will be covered word for word. Lastly, the attorney may state:

Q: Notify trial counsel if your opinion changes.

The impact of the deposition is usually negative on you. Holes in your story are exposed. Your background somehow seems inadequate. The other side has some valid arguments. That is the way it is. All you can do is attempt to minimize losses. Do not expect or try to score any points. Recall that the purpose of the deposition is to discredit and impeach you through inquisition. Be tough and resolute, and somehow smile and shake hands with your inquisitors when it is over. They are only doing their job.

THE JURY AND THE EXPERT WITNESS

Jury's Importance

The jury decides the case. They deliberate on the evidence presented and choose which version of the truth is correct. The jury is your audience, your constituency when you are on the witness stand. Your answers to questions must both be understood by the jury and have impact on them.

These simple realities are easily forgotten when engaged in combat under cross examination. But your examiner has not forgotten. With every thought and gesture oriented toward the jury, your examiner will be doing their utmost to distract you from the jury and lead you into positions where you will appear to be unintelligible or a person of ill temper or just plain different than them. The opposing attorney wants the jury to believe you are a hired gun.

The wedge the attorney creates between you and the jury is all to his or her advantage, starting when you direct your answers back to the attorney instead of to the jury. Responding to the examiner is an easy, natural thing to do; after all, they are who is asking the questions. But the attorney is not the jury. The attorney decides nothing. In responding to the attorney, you are ignoring the jury—and they know when they are being ignored. They know when you are talking over their heads. And they remember.

Who are these jurors?

Jury Composition

They are picked from the voter registration roles and are involuntarily requested to appear at the courthouse through a formal summons. Many potential jurors are reluctant to serve, particularly if the trial promises to be a long one. Unless they have a valid excuse, like young children to care for or some medical condition, the court will impanel them.

Once the jury is sworn in, jury watching begins. The size of the jury varies from six to twelve, with possibly several alternate jurors. Which jurors are paying attention, taking notes, shaking their heads? Jurors will begin to feel a bond with the attorneys. Which ones feel what bonds? Which juror looks like a leader, who may become the jury foreman?

Jury Response

Jury-watching can give important clues to when they get turned off or turned on by individual witnesses. Attorneys will look for clues to guide them as to how long they should interrogate witnesses and how high their amplification should be.

If your attorney wants you in the courtroom prior to your testimony, and you are allowed there, you need to watch the jury. Which jurors respond well to technical arguments? Which ones need help in understanding just the basics? How can your "pitch" be broad enough to encompass the understanding of the least educated juror while not boring the sophisticated ones?

Whether or not "the rule" of sequestration is invoked will determine your access to the courtroom. "The rule" states that witnesses may not be present in the courtroom while other witnesses are testifying. This prevents witnesses from making up stories that fit in with other testimony.

Attorneys may by prior agreement waive the rule for Experts. Because Experts testify only as to their opinions, they need to know everything about the case, including witness testimony. In many cases the rule is waived.

You are not allowed to speak with any juror or interact in any way. Sometimes Experts are asked to sit with their attorneys in the courtroom during the trial to help formulate questions. That is a bad practice, because it destroys the notion of an independent, objective Expert in the eyes of the jury. The Expert should remain in the spectator benches and have questions passed quietly to the lawyers during the proceedings or wait until a break to interact. Experts should blend into the courtroom without notice until called upon to testify.

When it is your turn to testify, you will confidently walk up to the witness stand and be sworn in. As your attorney asks questions, you turn to the jury to answer. Make eye contact with individual jurors. Establish rapport as quickly as possible and hold it. Look at your attorney when the question is asked, and swing to the jury to respond.

A negative impact can result from a direct examination when the attorney and the Expert act out a duet that is too perfect. The jury sees a rehearsed, scripted relationship that clearly shows which side the Expert is on. The testimony is discounted accordingly. Attorneys and Experts can get so wrapped up in themselves and their performance that the mutual congratulations are only melted away when the jury verdict arrives.

When cross examination begins, the jury's attention will be riveted on you and your examiner. He or she will, by the force of their personality, draw you into a one-on-one confrontation. They want you to completely ignore the jury and will do everything to demand your attention. Your counteraction is to look at the jury for every response. You must maintain a pleasing demeanor on the witness stand despite the histrionic behavior of your questioner. You are the Expert and you must play the role in an imperturbable manner.

The examiner wants you to get emotional and engage in debate. They want you to show your true colors as a mouthpiece for the other side. "Are you an objective Expert?" The jury is thinking, "How credible is he?" You have a right to get disturbed if the examiner asks an obviously unfair or prejudicial question. However, let your lawyer object to the question before you get upset. Keep your cool and maintain your integrity.

If your story is not getting through to the jury, they will show it on their faces. Hopefully you will have explained the case in its most elemental form to your grandmother or some other nontechnical person at some time before the trial. Use that storytelling method. Remember, you must tell the same story three different ways at three different times to get through to the average human being. Practice the three versions in advance and find ways to relate them in the testimony. The cross examiner may slip up at some point and give you an opening. Go for it. I remember a case in which the examiner asked me a question that I could best answer by going over a key exhibit again. I bounded down from the witness stand to the exhibit in front of the jury while the examiner was trying to restrain me. For the next five minutes I reiterated every major point made on direct from that exhibit while I stared eyeball to eyeball at the jurors seated five feet away.

Be animated and be direct. The jury wants to do their job, but they also want the trial over with as quickly as possible. They did not choose to be there. Be interesting so that they want to listen to you, and be brief to respect their situation. The side that drags out the case, or appears to, will pay later.

Everything counts. The jurors see it all—which side is more competent, more efficient, gets along with the judge better, knows where to find evidence, moves at a faster pace. In closely contested cases, the little things count. You, too, will cast a long shadow on the jury; understand this and use it.

Always be deferential to the judge. To the jury, the judge is God. They look up to the judge for complete divine guidance. The judge may ask you questions or give you instructions. Never argue or comply begrudgingly. The judge should also be divinity to you; show respect.

DIRECT EXAMINATION

Introduction

You will be introduced to the jury during the opening statement. Your attorney will say the jury will be meeting you and that you will be discussing the facts and offering opinions. Their expectations are raised to a high level. You had better deliver what your attorney has promised. Make sure beforehand that you agree with every word of the opening statement pertaining to your testimony.

Your attorney will build the case for the jury through the fact witnesses. Then your turn comes to weave it all together. Start looking at the jury naturally, person to person, to establish rapport. Your direct testimony will first establish your credentials. A good attorney will tie in your credentials with the case. The questioning may go as follows:

Q: Please state for the jury your full name.

Q: What is your business?

Q: Have you come to court to state your opinion?

This last question is the "tickler." Answer this question only with "yes." Do not state your opinion. Your attorney is establishing your reason for being at trial to pique the jury's interest.

Q: What are your credentials?

In the answer you tell the jury why they should pay attention to you.

> Q: What courses in college did you take that relate directly to the issue at hand?

> Q: How does such and such course relate to the cause of the problem here?

The lawyer in this line of questions is not just trying to show how smart you are, but why you are here and how your expertise directly bears on the issue. Finally, your attorney will tender you as an Expert in the relevant areas.

Voir Dire

At this point the opposing attorney has the opportunity to challenge your credentials. Most likely they will. Their primary strategy will be to limit your expertise. They also want to break up the flow of direct testimony and assert their authority. They will ask leading questions, such as:

> Q: Doctor, you do not claim expertise in the following areas, do you?

Or,

> Q: In your deposition of such and such date, you stated that you had no knowledge of _____, didn't you?

Be prepared to offer explanations of your expertise that will not limit the range of your testimony. For example, you may respond:

> A: Yes, I have no practical experience designing such systems, but it is well within my area of expertise to understand the theory.

Or,

Q: Yes, I have a professional license which allows me to
   practice in that area.

After your questioning, the judge will render a decision on
your expertise. You will be allowed to continue the testimony
unless some fatal flaw exists in your credentials.

Your lawyer continues with questions directed to your research and opinions.

Q: Doctor, we have reviewed your qualifications; based on
   your expertise and knowledge of the case, do you have
   an opinion to a reasonable degree of certainty?

A: Yes.

Q: What is your opinion?

A: [State your opinion.]

Next, the attorney will tie the individual elements of the case
together.

Q: Was _____ caused by _____?

Q: Do you have an opinion to a reasonable degree of certainty?

Your attorney will explore the basis of each opinion. In the
last part of the testimony, he or she will anticipate what the
other side will attack by exposing the weaknesses of his or her
client's case first, in the most sympathetic manner.

Q: Have you considered the following "weaknesses"?

Q: What effect does that fact have on your conclusion?

Q: Did you also consider _____?

Q: Are you familiar with the opinions of the other Expert?

Q: Do they cause you to change your opinion?

Q: Did you make any assumptions?

Q: How do those affect your opinion?

Forthrightness and Objectivity

The credibility of an Expert Witness must be consciously maintained in both direct as well as in cross examination. Many times credibility can suffer in the direct exam phase when the Expert's own lawyer is doing the questioning. Why? The Experts can easily come across as "homers," rooting for their own team.

Even though, practically speaking, he or she is employed by one side only, the Expert must show independence of thought to the jurors. The Expert must be perceived as a responsible, objective viewer of reality.

The jury knows that the lawyer is hired to be an advocate for one side. Yet the witness is supposed to be above that. If the witness looks like a hired gun, the testimony suffers. It is a paradoxical situation. On one hand, Experts will present opinions that reflect the viewpoints of the side that hired them; on the other hand, they must show independence of thought in coming up with those opinions.

The best way to show independence during direct examination is not to allow your attorney to take you beyond your bounds. Lawyers will push you as far as they can, to prove their position. It is up to you to keep from going beyond the limits of your expertise. You must draw the line. Don't be a mouthpiece for your lawyer. Fight your lawyer when they go beyond reasonable limits. The jury will appreciate seeing a measure of independence. Your attorney may or may not like it. If they understand what you are doing and it reflects positively on the jury, then your lawyer will appreciate your approach.

Remember that anything you say in open court or in deposition is public record. Your statements may stay with you for the rest of your career. That stretched statement you made five years ago can be thrown back in your face at any time, to your regret. Your honor and credibility are all you have to offer. You must protect them from even the hint of tarnish. Do not allow lawyers to push you beyond your limit—ever!

Credibility

How you answer questions gives important clues to the jury about your trustworthiness. If you don't know the answer to some question, the answer is simply, "I don't know." Unfortunately, many times Experts have a hard time expressing ignorance. They feel they must know it all. So instead of being direct, the answers may become:

A: I don't have any specific knowledge.

A: I may have.

A: I have no direct evidence.

A: I don't think I did.

A: I probably had.

A: I don't recall the details.

These hedging answers are smokescreens that are easily seen through and can only damage the witness. Avoid such responses at all times.

The following hypothetical situation can occur during direct examination.

Q: Doctor, as you previously stated, you believe the corrosion of the equipment resulted in greater vibrations and eventual fracture. Can corrosion be prevented by the addition of an inhibitor?

In this case, the witness is an Expert in vibrations with some knowledge of corrosion. The lawyer is attempting to draw him beyond his limited knowledge of corrosion to the area of corrosion inhibition. The question is seductive in that it begs for a simple "yes" answer. Let's suppose, however, that this Expert has never been exposed to corrosion inhibition techniques. A good response would be:

A: I can't address that question; it is beyond my area of expertise. However, I can safely and emphatically state that the corrosion was responsible for the vibrations and that if the corrosion had not occurred the machine would not have failed.

Two good things happened with this response. First, the Expert clearly drew the line on his expertise, showing that he was not just a mouthpiece for his attorney. Second, he clearly reemphasized his Expert technical evaluation. The jury can easily draw its own conclusions as to what would have happened, had the corrosion been inhibited. Such forthrightness wins big points.

CROSS EXAMINATION

Under Attack

How does it feel to be boiled in your own blood? That is one of the many emotions I have felt during the cross examination of the Expert Witness. The opposing attorney has the opportunity to question the validity of your opinions expressed during direct examination. They will also question the veracity of the witness—*you*. This is the opportunity the opposing attorney has been waiting for. The strategy is to impeach your testimony and destroy your credibility. They want to make you sound unbelievable to the jury.

The attorney has spent much of their professional life training to discredit you in cross examination. Beginning in law school, the give and take of sharp classroom debates between professors and students underpin the role of the adversarial approach in litigation. Television portrayals of courtroom scenes and newspaper headlines of trials only reinforce the drama of the battle between the attorney and witness.

From the perspective of the Expert Witness, litigation is an unfair battle fought in alien surroundings. Technical training consists mainly of the acquisition of knowledge under passive circumstances in the classroom, where few professors debate scientific or technical issues. "Here is the truth; understand it." On the other hand, "here are the facts; debate them," is the *modus operandi* for the lawyer.

Opposing Counsel's Strategies

The Expert tends to believe in a universal truth or scientific principle which—along with the facts—will make the case understandable. The opposing attorney believes that the facts are relative, depending on whose eyes are observing, and that the truths become evident as facts are explained in the proper light. Through cross examination, it is the obligation of the opposing attorney to shift the perception of truth. Your attorney believes in the same principles, and will apply them during cross examination of the opposing Expert.

The interpretation of facts—the Expert's role—is the contentious element. Unless the issue is on the cutting edge of knowledge, the scientific and technical principles are generally agreed to by both sides. Therefore, it is the duty of the opposing attorney to make you reexamine your understanding of the facts so that they can be interpreted more favorably for their side. They have three major techniques:

1. Initially, they will go over all facts agreed to by both sides. You will therefore be likely to start your cross examination by saying "yes" to most of the questions. Then the attorney will suddenly shift into the grey areas and skillfully push you into agreement in areas favorable to their side. Psychologically, you are predisposed to continue agreement with the attorney, having developed a nonthreatening relationship with the "yes" question-answer routine. You will be pushed to your limit with clever lines of questioning based

on your earlier answers. They will go for a series of damaging admissions as the questioning toughens.

2. They will confront you with any contradictions and errors in your work and testimony. The attorneys and their staff have thoroughly researched your professional life. Everything you have written on the subject at hand and previously stated under oath is potential material for the spotlight.

    You probably have written an Expert Report and given a deposition. This is especially inviting territory. Also, they will not hesitate to confront you with conflicting testimony from witnesses that have already testified. The battle here is on a high intellectual level, with every utterance you make being important.

    This battle is reminiscent of the Knights of the Roundtable in a jousting tournament, on horseback in their coats of armor. Time after time two riders approach each other with long poles and, at a full gallop, attempt to knock each other off their mount.

    In the courtroom, the jouster uses the long poles of mental acumen. Question-Answer: did anyone get knocked off the horse? Back for another round: Question-Answer. The battle is exhausting. When it is over, the opposition can turn to a third technique:

3. They will attack the Expert's credentials. This can be the bloodiest battle of all, since it is fought inside the territory of the Expert. Your credentials are being challenged as inadequate. And every Expert has weaknesses. An academic Expert is usually strong on theory and weak on experience. The practicing professional has the opposite problem: strong in experience, but weak on theory. The examiner will attempt to differentiate the Expert's experience from the factual situation of the case. Since every case is different, no one's experience exactly replicates the case. The attorney will try to make a big deal out of small differences. The attorney may attempt to show that their Expert is better than you.

This effort can have a high yield if the questions get under the Expert's skin. It can be hard not to take this personally. Keep in mind that it is the intent of the opposing attorney to have you "blow your cool." The game here is to expose the Expert as being sensitive to criticism and nonobjective in his or her opinions. Despite the natural psychological tendency to strike back, become defensive and/or argumentative, the Expert must be dignified and controlled under these tough circumstances. Once the examiner has pierced the veil of objectivity, they will exploit that opening you have allowed. The opposing attorney would like nothing better than to have a shouting match, discrediting the Expert. Remember that the opposing attorney has only one goal—to help their client. They will ask leading questions, those which suggest an answer.

"Yes or No" Answers

Cross examination need not be a purely defensive situation for the Expert, however. When the examiner asks a question, a potential opportunity exists for reinforcing points made under direct examination. The jury has the opportunity to reinforce their understanding of your opinions. Do not let the examiner cut you off or limit your responses. You may have to answer any question "Yes" or "No," but you also have the right to explain any answer. That explanation is your chance. And, if the question cannot be answered with a simple "Yes" or "No," say so. The lawyer may even try to cut you off with a "Thank you, that's all." You can insist on explaining your answer, and the lawyer must listen.

The Use of Hypotheticals

Lawyers love to use hypotheticals to question Expert Witnesses. The hypotheticals can clear up the differences and similarities between the Expert's opinion and those of the opponent's Expert. However, in the hands of a clever attorney, hy-

potheticals can be crafted to create confusion and gain agreement where there is none. Whenever the examiner poses a
hypothetical, red flags should be waving in your mind. There
is danger ahead.

The first duty of an Expert confronted with a hypothetical is
to ask for the assumptions, as follows:

Q: Sir, assume that the following facts are true: Facts A, B,
   C, and D.

A: But those are not the facts.

Q: Assume for the hypothetical that they are.

A: Okay.

At this point Experts have some choices: they can answer
with no response, as follows:

A: I need time to consider this hypothetical. It can't be responded to with a quick answer. It needs to be studied.

The examiner may at this point try to change the facts to
make you answer, or, if the examination is a lengthy one, ask
you to think about it over the break or overnight, and then
bring it up later.

Another way to respond is to alter the facts given to you, as
follows:

A: If Fact B was changed to Fact C, then my answer would
   be . . .

Or, you can answer the hypothetical directly. This is danger-
ous because the examiner loaded the hypothetical to benefit
their side.

What the examiner is doing is setting the stage for later on,
either with you or their own Expert, to show that the hypothetical situation with Facts A, B, C, and D is indeed the real
case, and that you misinterpreted these facts. And by your

responses, if you had clearly understood the fact situation, you too would have agreed with the opposing view. Just look at your responses to the hypothetical.

You must clearly indicate by your responses how the facts in the hypothetical are different from the facts you used to reach your opinions. Unless, of course, the facts are all the same. You are required—by law and in the eyes of the jury—to answer all questions put to you by the examiner. Evasion only helps your examiner and their side. Remember, you are the Expert.

An unforgettable moment for me involved the use of a hypothetical in the deposition of the opposing Expert in a recent case. He had vigorously resisted answering questions dealing with the actual fact situation by stating that he had not studied it in detail. No amount of questioning could find a crack in his resolve.

At a break in the proceedings the suggestion was made to pose the facts of the case as a hypothetical situation, since he was unwilling to reveal his opinions on the subject directly but might be induced to do so in a hypothetical situation. It worked! He sang like a bird. In his mind, the hypothetical situation was different from the case, and he felt he could freely answer. In reality, there was no difference. Only the word "hypothetical" separated the facts from fiction. And it was easily shown later that the hypothetical situation was indeed a mirror of the real situation.

Another valuable technique to use when confronted with a hypothetical is to request more detail about each part or to require more assumptions. For instance:

A: Fact B needs some time constraint.

Q: Okay, assume some value for a time constant.

A: In the case of infinite time, . . .

Time increments greater than zero alter the facts in the hypothetical to illustrate the points you made earlier in direct examination. The examiner will attempt to resist you:

Q: But these times are not realistic, are they?

A: Yes, they are. In fact, they show my point perfectly. [Explanation.]

With this show of resistance, the cross examiner may back off and wait until their own Expert can illustrate the points. Remember that hypotheticals represent danger and opportunity. Before testifying, go through potential hypotheticals in your mind and figure out how you will handle them.

A hypothetical question from an opposing attorney is only effective if your answer comes right after the question. Once the question has been repeated, reworded, and dissected, by the time you answer the jury will have forgotten the question or grown so weary that the point becomes meaningless.

How to Handle Questions About Your Fee

You have been paid and paid well for your work on the case. And you have been paid by one side in the litigation. The question is, if you have been paid by one side, how can you be objective? This is easy to answer. You are a well-respected Expert in your field and certainly would not compromise your high ideal for this case.

The opposing counsel will not directly confront you with the issue of payment and objectivity. Instead, the indirect approach will be used, as follows:

Q: How much have you been paid to testify?

An inflammatory question. The best answer is:

A: I have not been paid to testify. I charge a fee for my services based on the time I devote to the project.

Q: Okay, how much have you been paid for your services?

The question is usually stated in a sarcastic manner.

A: My consulting fee is _____, based on _____/hour for _____ hours. This is my regular fee for services, and since I provided services, I am remunerated.

You have fought off the examiner as well as possible for the time being.

The jury has heard what may sound to them like an exorbitant fee; but then again, they pay a premium price when they visit a medical doctor or other professional.

The examiner may hesitate at this time, knowing that his or her Expert may be subject to the same line of questioning. But expect the information about your fee to be followed up on. Approaches the examiner may use are shown in these examples:

Q: You mean you charged X thousands of dollars to perform this study?

A: Yes, the time also includes meetings, the reading and review of a stack of documents X feet high, experiments, literature reviews, and travel expenses.

Follow-up questions may be dramatically phrased.

Q: In this megabuck study, you found that _____?

Or

Q: You mean to tell me for those megabucks you _____?

Of course the examiner is trying to show how mercenary you were, and how much you charged for such minimal and flawed work. You are another living example of a high-priced hired gun willing to say anything for a price.

About all you want to do in these circumstances is maintain your composure and be dignified. Don't be defensive! You deserve to be paid for your services. This is not charity. The examiner is probably billing his or her client at a higher rate than you are. Let them play their games, and refuse to mud wrestle on the subject.

Sometimes at the end of a cross examination, the attorney will look at you incredulously and ask:

Q: You mean that you charged megabucks for these findings?

A: Yes.

By simply and dispassionately stating "Yes" you have maintained your dignity.

Protection by Your Attorney

The possibility is always present that you will get into trouble on cross examination. The examiner may be exploring areas you are not familiar with, or they may have found an embarrassing omission or mistake, or made you agree to something you shouldn't have, or caught you in a contradiction. Lawyers, by the nature of their training, sense when things are going wrong on the witness stand. The question is, how can your lawyer transmit his or her feelings to you, to avert a possible catastrophe?

If, during direct or cross examination, there are objections by either attorney, stop talking and wait for a ruling by the judge on the objection. You are not permitted to enter into the discussion on the merits of the objection, even if you have an opinion. That is for the lawyers only.

If your attorney objects to a question, one reason they may be doing it is to give you more time to consider your answer, or even to reconsider the direction in which you are going. This is particularly true if objections are being overruled by the judge. The attorney is sacrificing himself to protect you.

What can you do when your lawyer has indicated trouble in the making? Increase your resistance to the examiner. Possible resisting responses are:

A: Could you repeat the question?

A: Could you rephrase the question?

A: That question is too vague. Could you be more specific?

A: I don't understand the question.

A: The answer would require more study.

A: That's a difficult question to answer without more information.

A: That question demands that I make a whole series of
assumptions.

Make the examiner work. The examiner is a human being
and will get tired and worn down by a witness who makes
him/her work. That doesn't mean throw up roadblocks and
obstacles all the time. That is counterproductive; it alienates
you from the judge and jury. Pick your fights.

Testing Your Credentials

During cross examination, the questioner may again attack
your credentials.

Q: You are not an engineer, are you?

A: No, I'm a chemist, but I do chemical engineering work
all the time. I'm a practical engineer; let's put it that way.

Q: But you are not a registered P.E.?

A: No, I am not.

Q: Neither do you have an academic degree in engineering,
do you?

A: No, but we have people on the staff that have that type
of capability and academic degree, yes.

Q: But you personally don't have that?

A: No, I'm a chemist, right.

Q: You are not an Expert on the transport processes of advection, convection, and diffusion, are you?

A: No, I am not.

Q: Are you an Expert in mass transfer?

A: No, I'm not.

The witness was attempting to stretch his expertise beyond the bounds of his training in science to include engineering. The lawyer questioned the Expert's qualifications as an engineer by establishing that he had no engineering degree, was not licensed to practice engineering, and did not profess to know anything in the engineering areas of interest. The jury saw the witness attempt the "stretch" and later admit that he had gone too far. What a poor way to begin a relationship with the jury.

The witness would have been much better served through the following responses:

Q: You are not an engineer, are you?

A: No, I'm a chemist with knowledge of engineering, having worked for many years with engineers.

Q: But you are not an engineer, are you?

A: No.

Q: You are not an Expert, are you, on the transport processes of advection, convection, and diffusion?

A: These processes are within my general area of expertise as a scientist.

Q: But you are not an Expert in the area of engineering mass transfer?

A: No, only from a scientific standpoint.

The witness does not want his testimony to be so limited and constrained that he cannot testify openly on the subject. The jury will have only a fuzzy understanding of the difference between a scientist and an engineer. The witness has kept his options open by indicating a general understanding of the pertinent scientific principles, yet he forthrightly portrayed himself as a scientist. He has preserved his credibility with the jury.

One of the problems an Expert always faces is determining the breadth of their expertise. How general in the subject matter are you?

Is a heart surgeon competent to testify about the impact of a blood clot in the brain? Is a specialist on commodity transactions competent to testify on the buying and selling of stock? These are tough but arbitrary questions that only a judge and jury can answer.

The main question is whether the issue at hand is in your area of depth. If the case with the heart surgeon deals with the action of the heart and its impact on clotting in the brain, the surgeon should feel comfortable. However, if the main issue is clotting in the brain with the action of the heart being one of a number of suspected causes, then an Expert in blood disorders may be more appropriate.

In big cases requiring broad knowledge of a field, a generalist should be retained to give the "big picture," in addition to other Experts covering specifically related areas.

Judges generally are lenient in allowing the testimony of "Experts." They do not want to have a retrial based on error from not allowing a prepared witness to testify. Better to have the jury make the evaluation of the Expert's competency. A witness offered as an Expert who is weak will be subjected to rigorous cross examination on their competency for the jury to see.

The best way to handle the questioning of your breadth of expertise is as follows:

Q: Sir, you testified that your expertise was in mechanical vibration, not corrosion, isn't that true?

A: No. My main area is vibration, but I am qualified by law through my professional license and have knowledge about corrosion.

Q: But sir, you are not an Expert in corrosion, are you?

A: It is within my general area of expertise.

Q: Sir, please answer yes or no. You have no publications in the corrosion literature, do you?

A: No, but I have studied how corrosion affects vibration, and in that way I have become knowledgeable about the subject.

The Expert indicated very nicely how the subject of corrosion interfaced with his depth area of vibration. Also, by answering politely with an explanation each time, he deflected the age-old technique of lawyers attempting to restrict the witness to answer "yes" or "no." Some lawyers can get nasty at this juncture, as follows:

Q: Judge, please instruct the witness to answer the question "yes" or "no."

The Judge will instruct you for their own benefit to answer "yes" or "no." But the Judge will not instruct you to *restrict* your answer to "yes" or "no." You may add your explanation. Remember, your real audience is the jury, not the judge or opposing lawyer. You must explain to them why you answer the way you do. The lawyer has a line of questions on their yellow pad ready to trap you with "yes" or "no" responses. Let's see this in action:

Q: Your area of expertise is mechanical vibration, correct?

A: Yes.

Q: You have never written papers about corrosion specifically, have you?

A: No.

Q: You will consult with corrosion Experts when you have a problem in that area, won't you?

A: Yes.

Q: That's a normal thing, to consult with others in allied fields, isn't it?

A: Yes.

Q: You don't consider corrosion your area of expertise, do you?

A: No.

The lawyer has trapped this witness with a set of four questions that lead to the clincher. In the jurors' eyes, this Expert knows little about corrosion, when in fact he may know much. The "yes" or "no" answers got him in big trouble.

Furthermore, the lawyer has set a rhythm and established a relationship of dominance and submission with the witness. This does not bode well for the witness.

Now that the witness has been compromised, he will be subjected to further limiting of his expertise:

Q: Your work has been largely theoretical, hasn't it?

A: Yes, but I have used the data of other researchers to confirm my theories.

Q: But, you have never conducted vibration tests yourself, have you?

A: No.

Q: And you never tested the piece of equipment that is the subject of this lawsuit, have you?

A: No.

Q: Did you ask to have the machine tested?

A: No.

Q: Did you know you could have tested the machine?

A: No.

Q: So your only work was a computer analysis of the vibration, correct?

A: Well, it wasn't the only thing I did.

Q: But it was the work upon which you based your conclusions, wasn't it?

A: Yes.

Q: And isn't it true that with computers, it's "garbage in, garbage out," that the validity of the model is based on the assumptions you have made?

A: Yes, the model is based on my assumptions, but I don't consider it garbage in, garbage out.

The attorney is having fun with this witness. First, he got him to agree that he is purely a theoretician with no practical expertise. That causes further separation between the jury and the witness. Second, the attorney belittled computer models by planting in the minds of the jury the image of a computer as garbage. And since most jurors have had some kind of computer foul-up in their lives, they can relate to that.

This Expert, as all Experts must, has to fight the lawyer on his intellectual premises. The battle is for the minds of the jurors, and secondarily, the Judge. The following responses will result in a vastly different outcome:

Q: Your work has been largely theoretical, hasn't it?

A: Yes, I develop theories to explain what happens in the real world. I take information on vibrations from other researchers who test the equipment, and use this data to explain what happened. That is what I did in this case: I took data and used my theories to explain what happened.

Q: But, you have never conducted vibration tests yourself, have you?

A: No, but I know exactly what these tests are about, how they are run, and what the data indicates. I collaborate with the experimentalists all the time. My work is to explain what happens through theory.

Q: And you never tested the piece of equipment subject to this lawsuit, have you?

A: No, that was not necessary (or possible). I took the data that had been generated and analyzed it.

Q: Did you ask to have the machine tested?

A: No, I didn't need to. I had sufficient data to figure out what happened.

Q: Did you know you could have tested the machine?

A: No, the machine did not have to be tested further. I had the data I needed.

Q: So your only work was a computer analysis of the vibration, correct?

A: No, I conducted a literature review comparing my results with other researchers in the field. I also did sensitivity analyses to check on the precision of my assumptions. And lastly, I discussed my findings with other Experts in the field.

Q: And isn't it true that with computers, it is "garbage in, garbage out"?

A: No, as I mentioned, I carefully evaluated all the imprints to my model and then checked the results with others in the field. I stand behind my conclusions.

With these answers, the Expert seized the opportunity to refix in the jurors' minds the care he took to arrive at the conclusions. The attorney is questioning whether to go on with the cross examination from his prepared questions on the yellow legal pad, or to bail out at this time. The counterpunches by the Expert are taking a heavy toll.

Leading You Out of Your Expertise

If the opposing attorney cannot get you to limit your expertise to their advantage, they may take the opposite approach and gently allow you to stretch yourself beyond your limits so that later they can saw off the branch on which you have stranded yourself. Sometimes even your own attorney will want to stretch your expertise beyond the comfort zone. In either case, you must draw the line.

An example of stretching:

Q: As an engineer, you have indicated that the toxic chemical will come in contact with humans, correct?

A: Yes.

Q: You also indicated that the concentration will be very low, correct?

A: Yes.

Q: Well, at that low concentration, the toxic chemical could not have any impact on human beings, could it?

A: Yes, it could accumulate in the body and build up to such a level as to cause harm.

Wrong answer! Unless you are a toxicologist, or medical doctor, you do not have the qualifications to expound on chemical effects in the human body. You have exceeded your limits, even though the statement you made may be true and you know enough from the background investigation to say it. The correct answer is:

A: I am not a toxicologist and cannot answer that question. However, the health standard limit for human exposure is _____, and that value was exceeded.

The jury will appreciate that you know your limits as an Expert, yet you can still testify to some recognized standard that professionals go by in evaluating effects.

The wrong answer would have been followed by this line of questions:

Q: Where in the body does the chemical accumulate?

A: I believe it is in the liver.

Q: And what effect does it have on the liver?

A: It causes problems with the functioning of the liver.

Q: Be more specific. What problems?

A: I don't know specifically.

Now the lawyer has you professing ignorance. He may go on.

Q: What is a toxicologist?

A: One who studies the effects of chemicals on living things.

Q: You are not a toxicologist, are you?

A: No.

Q: You have no degrees in toxicology, do you?

A: No.

Q: You are not a medical doctor, are you?

A: No.

Q: All you know is what you read about the toxic chemical's effects on humans?

A: Yes

Q: Your honor, I move that the witness's answers on the

toxic chemical's effects be stricken from the record, and the jury be so instructed.

Judge: The jury is instructed to ignore those comments.

What a devastating line of questions! The Expert looks like a fool who is unable to discern between what he knows and doesn't know. Later, when a real toxicologist shows up, the lawyer will continue the contrast.

Remember, the diminution of opposing witness testimony is one of the lawyer's most important challenges; and they spend much of their working lives preparing to do just that.

Use of Inflammatory Words

"Them is fightin' words"—those that arouse your passions and place you in a fighting mood. But you must not allow that. Remember, you are the Expert, and you are above all that.

Q: Doctor, you mean you spent all this time and effort on the case and totally ignored this important piece of evidence?

Or

Q: Sir, you mean to tell me that you disagree with all the other distinguished Experts in the field on this issue?

Or

Q: This mistake you made . . . how many more mistakes are there in your work?

Or

Q: Could it be, Mister, that you are flat wrong?

Or, turning up the heat,

Q: You just blew it, didn't you, Doctor?

Or

Q: Sir, you really don't know what you are talking about, because you didn't bother to check?

Or

Q: You have been criticized by your peers before on this issue, haven't you?

It's hard to resist fighting back, but you must. As serenely as possible, indicate the flaws in the logic of the question and why things are the way they are. For example:

Q: You just blew it, didn't you, Doctor?

A: No. There was a minor error, but it had a negligible effect on the results, for the following reasons.

If you had really blown it, the examiner would have strung together a series of withering questions to put an airtight seal on your blunder.

Another common tactic by the cross examiner is to proceed to an easel and list the things you didn't do or the reports you didn't read, or the documents you didn't use. The headline will be:

Items Not Considered

The examiner will bring up as many items as possible, intimating their extreme importance. Your obvious duty is to indicate why those items were not important to consider. If you cannot repair the damage sufficiently during cross examination, there is always redirect.

Open-Ended Questions

Most of the time the cross examiner will have prepared the complete list of questions they want you to answer. Lawyers hate surprises, so the questions are carefully honed to elicit the appropriate response. If your deposition was taken, you can expect the best questions (worst from your point of view) will be reviewed and asked again. And if you deviate from the earlier responses, you will be made aware immediately of that deviation and asked to explain it.

Your explanation may be quite reasonable, such as:

Q: Doctor, your response is different in the courtroom today than it was on such and such a date, isn't it?

A: Yes, I have further researched the matter and found [that it was in error; or that there are newly uncovered facts which are . . .]

A trick commonly used by cross examiners is to take the question and answer out of context and try to relate it to the line of questioning. For example:

Q: You previously stated, on page ___ of the deposition, and I quote: "The drinking water was not contaminated by dimethydoubledeath (000)." Now you're saying that it is, aren't you?

A: You are taking that quote out of context. Those questions involved the time period before the alleged poisoning [or after, or in another situation].

You are always entitled to see the transcript and review the passage in question. Clearly differentiate how the statement was used out of context. Have the examiner read the questions and answers prior to the pertinent question to show the judge and jury the examiner's duplicity.

One of the toughest challenges to my credibility occurred during a trial where the cross examiner pointed out that about half of the subject matter about which I was proposing to testify to I had claimed no expertise in at deposition. The questioning at the deposition went as follows:

Q: You have never designed a cooling tower, have you?

A: No.

Q: Doctor, we understand you are an Expert in cooling water treatment, but you are not also claiming to be an Expert in the design of cooling towers too, are you?

A: No.

Then, during the trial, the questioning went:

Q: Dr. Matson, you are not an Expert in cooling towers, are you?

A: Well, I am familiar with cooling towers. It is within my general area of expertise.

Q: But, sir, in your prior deposition you stated you are not an Expert.

A: I did?

Q: Yes, let me read the Q&A: Dr. Matson, you are not an Expert in cooling towers, are you?

After a long pause . . .

A: It is true I am not an Expert in the nuts and bolts of the design of cooling towers. I am very familiar with the principles of design, the theories of operation. I have to understand those things to deal with the water treatment.

There was no follow-up.

Impromptu Questions

The examiner will deviate from their yellow pad of questions when the answers they are looking for are not forthcoming, or some unexpected response comes, or if they have some inspiration occur to them on the spur of the moment. Some lawyers rely more on their intuition than on the yellow pad. When the lawyer deviates from prepared questions, the odds become more even in the intellectual battle. The examiner must formulate a question that not only will not be objected to, but will provide the desired response. That is when the questions become more open-ended and the witness has a better opportunity to respond. There is also some risk, however. The examiner would not deviate from the prepared text unless they felt there were fertile grounds to explore. The questioner is attempting to take the witness down some path. The witness has to make a decision to go down that path or to try to divert the questioner.

An experienced Expert will know when and how to draw the questioner in the desired direction. Usually the amount of resistance put up by the witness will be seen by the questioner as a cue of how sensitive the area is. For example:

Q: Sir, your theory of blood clotting does not apply to people on medication, does it?

A: It might.

Normally this would be a poor response to a question. It shows the Expert to be possibly equivocal.

A: What do you mean, "it might"?

This is the open-ended question you have been waiting for. Begin the monologue.

Watch for clues as to when the examiner leaves the prepared question list. Watch to see if they look down at the pad. Do they formulate questions and then change their mind? Lawyers will always prefer to ask leading questions in the form of statements, like:

Q: This is the way it is, isn't it?

Q: The coat was red, wasn't it?

Q: You will agree that the red color was due to an organic dye, correct?

Q: Organic dyes are toxic, aren't they, Sir?

But once lawyers start deviating, the questions become:

Q: That coat contained organic dyes that were toxic and were ingested by the infant, correct?

This is an example of a compound question that your attorney will object to, and the judge will ask the questioner to rephrase.

Q: Which dyes can be toxic?

Now the examiner is asking questions rather than making statements. The battle of the minds is on, and you have an opportunity to go for it.

Agreeing with the Opposition

The other way an Expert is coopted in the courtroom is by having the Expert agree with the opponent's version of the situation. Agreement is usually gained through psychological tactics that first create some bond between the questioner and the witness. A series of questions follows which leads to some agreement that the opponent's view has merit or plausibility.

A slick attorney knows that it is practically impossible to "turn" an Expert, that is, to have them change their mind during the questioning and agree fully with the opposition. However, seeking agreement that the opposition's theory is a possible explanation is a strategy that will invariably be used against the Expert.

It is human nature to find common ground when talking with another person. This holds true in the courtroom. The lawyer will exploit this to begin:

Q: Good morning, Dr. _____. My name is _____. We have met previously, haven't we?

A: Yes.

What will follow is a series of apparently innocuous questions to develop rapport. Questions are usually of a general nature on background, resulting in easy answers. No stress.

Then without any change in rhythm, the questions will evolve to:

Q: You will agree, won't you, that there can be various theories to explain what happened?

A: [Be careful, this is the loaded question: A "Yes" answer leads to:]

Q: And you will agree, won't you, that other qualified Experts may have different opinions of what happened?

A: [A "Yes" answer compounds the problem, as follows:]

Q: You are familiar with Dr. _____'s [the opposing Expert's] reputation and work, aren't you?

A: Yes.

Q: He is a man with an excellent reputation in the field, isn't he?

A: Yes.

The smart attorney will stop here. He has established that there may be more than one explanation and that his Expert is qualified to provide it. The attorney may go on, or return to a line of questions about the various theories, laying groundwork to attack yours.

Attorneys will not go for the kill unless that pathway is obviously open. They know that you are not on the witness stand to prove their case. They only want to neutralize you and your version of reality. You have helped them considerably. They want to present their ultimate conclusions in their closing argument to the jury.

Be alert to any question that is predicated by "You would agree . . ." Try these responses:

Q: You would agree, that other qualified Experts may have different opinions of what happened?

A: No, if they worked on the facts and understood the situation as well as I do, they would certainly agree with me.

Wow! You put an end to that line of questioning! You also put the burden on the other witness to prove that they have the same degree of understanding that you do.

Cross examination is over, but your testimony may not be over. Your attorney has the opportunity to ask questions on redirect, which may be followed by recross. Then the judge may ask questions. Finally, you can step down.

7

Communication

PSYCHOLOGICAL FACTORS

In this chapter we deal with the intangibles of communications, which are as important as the words you use. Remember that your testimony has two main purposes:

1. To provide a record for appeal

2. To convince the jury

Words that are transcribed and preserved become the trial record for the appellate process. The appeals court doesn't have the ability to feel the pulse of the courtroom or understand the emotional impact of your testimony. The jury does.

How you give your testimony, and the picture it paints in the individual juror's mind, is as important as what you say. The jury may listen to days and weeks of testimony. What they remember and carry with them into the jury deliberation room decides the case. Ninety percent of what they hear is forgotten. You want that remaining 10 percent to be the important points you make. How can you do that?

Storytelling

You need to leave the jury with a logical story they will not forget. It needs to be a consistent picture that is intellectually and emotionally anchored. They must feel and see, as well as understand, your contentions.

Simplify your testimony to, at the most, two or three main points. You must decide on one overall story or picture, with several supporting snapshots—more than that and the jury will become confused as to what your main points were.

For example, in a case involving engineering malpractice, a firm was being sued that had been owned by the founder, who later sold the firm and took semiretirement. During the transition, the firm took on more work than they could handle, and problems arose. The plaintiff had to choose what the main focus of the case should be. They could have shown that some of the people were incompetent, or that even though they were competent, they made serious technical errors. Or that because of the overwork, controls on quality diminished and the work became sloppy. Or that they were too inexperienced to work a big job and should not have taken it on. The latter version was chosen as the major focus.

In another case of engineering malpractice, the defendant established a clear vision. The case involved the design of an environmental process which did not function as specified. The vision presented was that the plant operated the system negligently and they were responsible for the problems. In the Expert testimony, that point was hammered home time and time again. And it worked. They won.

Most important, the story must be rooted in reality. If negligence is the main contention, the witnesses for the other side should be made to appear to have possibly been negligent, to have shown by past behavior that they have made mistakes in judgment or not fully understood what they were doing. A jury will not be convinced of negligent operation if those witnesses appear to be solid citizens who are eternally vigilant. If your contention is that the opposition is overworked and sloppy, their witnesses should have that appearance. Unrealistic visions create cognitive dissonance in the jurors' minds and hurt your case.

Emotions

To anchor your testimony in the emotions of the jurors, you must express some emotion in your testimony. You are not just a mechanical robot programmed to say what was noted in the prepared script. You feel more strongly about certain things than others. Those feelings need to surface and be shown.

This does not mean some emotional outburst such as anger or tears. It does mean appropriate voice inflection, bearing, and facial expression. Make your main points with strength of conviction; use hand gestures. "Yes—I am completely convinced that my theory is valid," stated while pressing forward with a clenched fist and determined expression, can strongly communicate your sincerity and belief.

There is nothing fake about showing emotion to buttress your main points. It alerts the jury that something important is being said and they had better pay attention.

The overuse of emotion can be detrimental, however, as impact is lost. The jury quickly tires of being brought to the edge of their seats for every minor point and reaction. Save yourself for the clinchers.

Emotion also means strong eye contact with the jurors. Draw them into your testimony and hold them there. Scan their faces while giving testimony. Recognize and respond to the puzzled looks on their faces. You are there to create understanding. Along with understanding, create empathy. Think of yourself as a tribal chieftain sitting in a circle describing an event to them. They want to understand you and they will be attentive if you remain attentive to them.

Leadership Without Primacy

As the leader of the tribe, your role is to develop a common understanding of events. You interpret the complex and make it simple. Never talk down. Events and theories are described at various levels of abstraction so that you make lists on the minds of jurors, of whom some think abstractly and others sense concretely.

Repeat what you say in different ways. For example, you may describe a chemical reaction in terms of charged ions that are attracted to each other through electrical forces. Alternatively, you create a metaphor of this phenomenon by describing the reaction as a marriage between a female who has a certain affinity for a male. You carry the metaphor through to arrive at the vision you are sponsoring.

The real connections are nonverbal. People quickly develop beliefs that are then backed up with evidence and words. How do you communicate nonverbally? Be aware of how you come across, of what kind of first impression you make. Perception becomes reality in the courtroom. You might even consider videotaping yourself in a practice cross examination session.

A "truthful" Expert maintains a rapport with the jury through eye contact. His or her demeanor is relaxed and confident. Speech patterns are slow, in low tones and measured words. Use a directness and openness to inspire the jury to listen. And raise your voice only to make important points. The jury feels your sincerity in the gut, and your words confirm their belief in you.

VISUALIZATION

Painting pictures with words and emotions is the most important role of the Expert. The jury must be left with a clear vision of your case that sticks with them throughout their deliberations. Visualizing means seeing, using the eyes and the mind, to picture your situation.

Visualizing is your most difficult task. You must take technically complex issues and break them down into components easily understood by the jury. To do this, you need to consider how the technical issue relates to the common symbology of mankind. A material undergoing corrosion needs to be explained in terms of a rusty nail. A surgical procedure may be explained as being like sewing a fine dress. The shifting soils beneath a tall building may have characteristics likened to a sand castle on a beach. Electrons flowing through a semiconductor may act like rats in a maze.

We grow up in a culture of common symbols (or symbology). We fall from bicycles, eat spinach, buy clothes, raise children, have a series of common experiences that bind the culture together. Your responsibility is to find the common symbols that illustrate those points in the simplest possible way.

But you must be careful. The opposition will be looking at those common symbols to use to their own advantage. I recall a case where a piece of equipment was installed to prevent corrosion in a process. It didn't. But there was no corrosion in the piece of equipment itself. The symbol we developed was that of a furnace installed in a house that just kept itself hot but did nothing for the house. The problem with that metaphor was that the opposition could counterattack, claiming the house was not built properly. We did not want to get into that so we did not use the symbol of the furnace.

In furtherance of visualization, any tangible object has significance. Try to bring technology into the courtroom. Show the piece of equipment in contention. If possible, take it apart and demonstrate your points. If that is not possible, consider a computer-generated model, an illustration, or a videotape. Compare what you bring in to some familiar, common object. An industrial boiler may become a teapot. A chemical reaction is like a pressure cooker.

Run an actual or simulated experiment, if possible. Make sure the experiment works the way you planned. Nothing is worse than a bungled experiment on the witness stand. Everyone is embarrassed, most of all your client.

Use words (adjectives) that bring vision to life in your oral testimony. Steel does not just lose strength through the erosive action of an abrasive substance, it *becomes weak* as the *sandpaper-like* material *eats away* at the metal. State the observations in technical terms, but then follow up with the vision in common symbols.

Each step in the visioning process leads the jury to inescapable conclusions. Pieces of information are linked together; the case develops like the unfolding of a complex plot, illustrating and fixing pictures in the jurors' minds forever. You are the producer, director, and main actor. The audience awaits your performance.

CONCLUSION

Show your humanity. Be who you are, and tie yourself to the human beings in the jury box. You are of the same flesh and blood. You have some specialized knowledge you want to share with them. That is your reason for being.

Effectiveness in the courtroom demands all of you and even more. Become aware of every level of your communication. The words you use, the gestures you make, the way you sit, the clothes you wear, are a composite of yourself. Developing awareness requires motivation and diligence. You have this determination within you, however, acquired while building the technical skills that qualify you as an Expert.

Section II

Case Studies

8

The Engineer's Nightmare:
A Case Study

THE FACTS

The Complaint

Another day in the life of Diane Gore, Attorney at Law. On her desk is the Complaint, the formal legal document that commences a lawsuit. Negotiations between Otto Industries and Skaff Engineering had broken down in an angry show of emotions. This is the next step. The slow process begins for her client, Otto Industries, seeking remedy through the American legal system.

On the cover:

In the Court of Common Pleas: Civil Division
Otto Industries, Plaintiff vs. Skaff Engineering, Defendant

Later that day, the Complaint is delivered to Bob Skaff, who is not surprised. He knew Otto Industries was taking legal action. What a nightmare, to be sued by your own customer as a result of their own, unreasonable behavior! Once the largest and most prized contract in his company, the turnkey job for Otto Industries had turned into a disaster in the field. Even before the job had been completed, finger pointing had gone

103

back and forth. After Skaff Engineering left the jobsite, the acrimony only increased. Embittered, both sides broke off negotiations and consulted their attorneys. The long, civilized warfare of legal attrition was to begin.

Engineers and technical professionals on both sides were to be embroiled in a contest that they were unprepared for, fought on a legal battleground that was alien to them, and dependent on warriors who spoke a foreign language—their attorneys.

Background

Three years earlier, Bob Skaff could not have been happier. Otto Industries had contacted him about a big job in his specialty, the design and construction of a wet scrubber system for cleaning particles out of waste gas from a boiler.

The technology was straightforward. A wet scrubber is nothing more than a large cylindrical tank with water sprays at the top. (See Figure 1.) The gas enters at the bottom and is mixed with the fine water spray. Water contact with the particles creates larger, heavier masses that fall to the bottom of the tank. The particle-laden water is shunted to a clarifier where the particles settle to the bottom and are removed as sludge. The cleansed water is then recycled to the tank.

This process was standard technology in the scrubber industry. What made Skaff Engineering different was his company's innovative use of special chemicals to prevent the formation of scale in the scrubber equipment.

Flue gas not only contained particles, but, depending on the type of coal burned in the boiler, calcium particles and sulfur dioxide as well. When water was sprayed in the gas, the sulfur dioxide was adsorbed and converted to sulfate that reacted with the calcium to form a tough, adherent scale called calcium sulfate.

White chalky scale would coat the walls of the scrubber and the water treatment equipment, causing operational difficulties and shutdowns. It was a real problem.

The standard method of coping with scale was the "seed-slurry process." Large concentrations of particles called "seed"

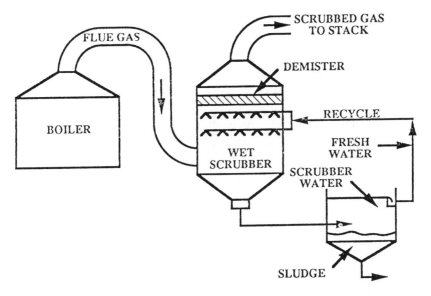

Figure 1. Schematic of a wet scrubber system for removing particles from waste gas from a boiler.

were put into the water as a slurry, so that some of the reaction between the calcium and sulfate would occur on the seed particles rather than on the equipment. This slurry process was not totally effective because scale still built up, though at a slower rate, and pumping the slurry was difficult. But it was better than nothing.

Bob Skaff had the magic bullet, he thought, in a special chemical formulation he called Dynamo. It interfered with the reaction between the calcium and sulfate so that no scale was produced. His lab had pioneered the use of scale inhibitors in wet scrubbers and he was proud of his new line of Dynamo chemicals, guaranteed to prevent scale in every application.

The chemical formulations used in Dynamo could not be patented since they were used in other markets, but Skaff was the only one with proprietary knowledge in the wet scrubber field.

When Otto Industries called Skaff Engineering, inquiring about his technology, Bob Skaff was confident that Dynamo would be effective. He had never done work for the coal-fired boiler industry but, after all, he had designed and installed over 200 scrubbers in all sorts of applications.

When Bob Skaff went to Otto Industries to discuss their inquiry, Chief Engineer Hamilton Burger greeted him and briefed him on a very pressing situation. The Environmental Protection Agency had recently negotiated a consent order with Otto. The order stated that Otto Industries had to comply with the EPA standards of 0.05 micrograms/cubic meter (0.05 $\mu g/m^3$) within two years or the boiler had to be shut down. Otto commissioned a feasibility study and found it would cost roughly five million dollars to install a conventional wet scrubber system. At a technical conference they had heard of Dynamo and wondered if it was an appropriate and less expensive technology. They were not experts in wet scrubbers and wanted a company who would perform a turnkey job. Was Skaff interested?

Of course Skaff was interested. The two-year time frame was very tight. There would be no time to run a pilot plant or do much laboratory test work. But Dynamo had worked in every application so far, even in other coal-burning applications. Bob Skaff went back to his office and sketched out a process flow diagram based on the volume of flow and particle loading information given to him by Chief Engineer Burger in their meeting. Bob Skaff estimated a turnkey price of three million dollars; Skaff Engineering would complete the job within the two-year period.

In the contract negotiations that followed, Skaff submitted a warranty which stated that the wet scrubber would at all times meet the 0.05 $\mu g/m^3$ dust standard, and further warranted that the water *"would not cause adherent scaling in the operating equipment, provided that the system is operated in accordance with Skaff instructions. Skaff limits its financial liability to the price of the contract (three million dollars)."*

Dynamo was effective; Skaff Engineering was willing to put it on the line. The contract was signed and the work commenced.

First thing, Skaff requested and received from Burger the

historical data on the flue gas volume and particle loading that had been taken by Otto Industries on a monthly basis for EPA compliance. This information could be the basis for the process design. With the tight time schedule, time could be saved by not performing any lab studies.

The process design was completed in three months—a record time. Skaff submitted the documents for Otto Industries to review, followed by a meeting at Burger's office. Burger had sent the drawings to his staff engineers, who responded with many questions about the mechanical details of the system. They wanted the new system to be fully integrated within the existing plant, and suggested several changes. For example, Otto Industries engineers wanted the control panel to be integrated into the existing control room, rather than located at the site of the wet scrubber.

Skaff was irritated by the volume of potential changes suggested by Burger and his staff. This job was on a tight timetable; making changes slowed everything down. The contract penalties ($5,000/day) were too severe to delay other work until changes were finalized. Skaff decided to order the major equipment immediately. During the construction phase it would be integrated with the changes that would have to be made to satisfy Burger. Skaff knew his engineers could do it. They always had before.

In the following weeks Bob Skaff became increasingly irritated at the bureaucratic handling of the changes by Burger's engineers. The nitpicking continued without full resolution. Otto Industries engineers felt frustrated, too, with the seeming lack of concern for details by Skaff Engineering.

Burger raised the question of the Dynamo chemical. How effective was it in this application? Skaff refused to answer Burger, reciting the clause in the contract under "Proprietary Technology": "*The information on the Dynamo chemical is considered proprietary.*" Skaff had similar problems with Burger when asking about the details of the boiler operation. Certain aspects could not be revealed.

The professional atmosphere between the two companies cooled to an icy, frozen state. Informal communications ceased;

instead, there was only the minimum interaction required to do the job. Routine shop drawings and construction sketches routed to Hamilton Burger were returned bleeding with red ink. Obviously, this project would be finished only at the last moment. Skaff reflected on the difficulty of working for a large company. Everything had to be submitted to a huge bureaucracy that chewed up his work and spit it out like a paper shredder.

Burger was concerned that Skaff and his staff were not sensitive to the problems of integrating the system into Otto Industries' plant. The Skaff engineers were too casual, indicating adjustments could be made in the field if problems arose. Burger knew better. In a large company like Otto Industries, changes had to be written down; operations went by the book. The system had to be designed correctly on paper if it was to be properly integrated, and he was having a rough time making Skaff understand that.

As construction proceeded, tempers flared between the Otto field engineer and Skaff's construction superintendent. The Otto field engineer questioned decisions that Skaff's people were making; he reported back to Burger that the construction work was being sloppily performed and the tie-ins were not being made in accordance with the drawings. At one meeting, the shouting rose to such a high level that both sides walked out.

The work continued. Four months before startup, Otto Industries operators went through a one-week training session with Skaff's engineers. Skaff's people were disappointed with the quality of the operators, but did not voice their concerns to Burger.

One month before startup, a thorough walk-through was conducted by Otto engineers. The design on paper had been transformed into a maze of lines, pumps, and tanks cramped into a dingy steel building. Hamilton Burger was frustrated, concerned, and disgusted. Bob Skaff knew the startup phase would be rough. The signs were foreboding.

Startup

The valves were turned on. Water sprayed into the scrubber as operation commenced. Almost immediately, dirty water from the scrubber overflowed the first basin, the clarifier. Someone had forgotten to open the valve on the discharge line from the clarifier. "A normal startup problem; nothing goes right for the first several weeks until the system is debugged," thought Skaff.

And problems there were. The wall in the sludge holding tank ripped loose and collapsed. Pressure from the wet well discharge pumps after the clarifier caused the pipe to rupture. Discharge lines from the wet scrubber leaked, spraying hot water and dirt into the work area, causing operators to run away in order to avoid being sprayed.

Then, as these problems were licked, the flue gas pressure exiting the scrubbers increased. This meant less air could be forced into the production boiler. Production of steam from the boiler decreased dramatically. The unit had to be shut down. Upon entering the scrubber, Skaff and Burger noticed scale on the demisters (honeycombed plastic that the existing air must flow through) installed to remove water droplets. Scale had caused the back pressure problem on the boiler, by restricting the openings in the demister. "Why didn't Dynamo work?" complained Burger. Skaff shrugged his shoulders, puzzled. "Dynamo has worked in every other installation we've designed. Something must be wrong with the equipment."

For the next month, Skaff's engineers worked overtime to solve the problem. The scrubber continued to scale up. Finally, one of Skaff's engineers suggested they temporarily add a seed sludge to the system and operate that way until a solution was found. Everybody agreed to this temporary fixup.

The fixup solved the scaling problem, but caused an operational nightmare. Pumps cavitated, lines plugged, valve seats wore out in a day, pump linings wore out in hours. This operational state could not continue. With no solution in sight, Burger sought permission to remove Skaff Engineers from the job-site. Approval was quickly given. So, four months after com-missioning of the equipment, Skaff was off the job. Otto Indus-tries still owed him $150,000—the 5% retainage from the job. Skaff still did not know what caused the failure of Dynamo; maybe it had something to do with the equipment and water chemistry, but he had no definitive answers.

Burger was angry. Skaff had represented his firm as a spe-cialist in this field, with expert knowledge and proprietary chemicals. Nothing worked. Burger was left with a jumbled mass of equipment that worked sporadically. Furthermore, EPA was breathing down his neck for him to meet the air emission standards, and he was out of time.

Over the next six months contractors were called in by Bur-ger to redo the pumps and piping and make the system oper-ate. A conventional recirculating seed slurry system was in-stalled, at a cost of one million dollars. With modifications, the system now worked. But it had to be shut down every six months to remove the scale—the same scale that Dynamo was supposed to prevent.

Negotiations began to settle the disputed claims between Otto Industries and Skaff Engineering. Otto Industries de-manded the million dollars they claimed it took to get the sys-tem running. Skaff Engineers stated they had not been given enough time to get the system fully functional, and that the operators were at fault for the system's lousy performance. Several times operators had allowed the Dynamo chemical tanks to run dry, and that was when the scaling problem oc-curred. If chemicals had always been in the system, no scaling would have occurred. Skaff offered to give up their claim for the $150,000 retainage just to get out of the imbroglio. Skaff and Burger talked about compromising at a half million dollars when Otto Industries abruptly ended the negotiations. Burger had had enough. "We'll let the lawyers decide this one," he commented as he left the room. Skaff yelled back, "See you in court, you _____!"

THE LAWSUIT

The Pleading

One year after startup, three years after the contract was signed, this formal legal document sat forebodingly on Skaff's desk. He had known it was coming, like a letter from the Internal Revenue Service, and he did not want to look inside. But he had to.

PLEADING

COMPLAINT

And now comes the plaintiff, Otto Industries (hereinafter, OI) who by its attorneys brings this complaint against defendants Skaff, for a cause of action of which the following is a statement:

IN ASSUMPSIT

Count 1

1. As a result of negotiations that previously took place between the parties, plaintiff OI entered into a contract with Skaff.

2. The contract included the following performance warrantee by Skaff that was a part of the consideration and benefit for which OI specifically bargained: "Skaff warrants that the water will not result in adherent scale to the operating equipment, provided the system is operated in accordance with Skaff instructors."

3. Skaff had the responsibility to design and construct the system and to ensure proper interface and compatibility with OI's boiler system.

4. OI operated the system in accordance with Skaff instructions.

5. Three weeks after the system was placed into operation an increase in gas pressure in the scrubber resulted in a shutdown of the boiler. An inspection of the operating system revealed adherent scale in the demisters.

6. The scale was removed. To prevent recurrence of scale, Skaff recommended that OI add seed slurry to the operating system.

7. The recommendation was implemented immediately. Thereafter, the abrasive properties of the slurry destroyed the linings of the pumps, valves, and lines, causing frequent shutdowns of the system.

8. Although OI advised Skaff repeatedly of the situation, Skaff failed to satisfactorily remedy the problem, which constituted a breach of warranty and a material breach of contract.

WHEREFORE, Plaintiff OI demands judgment against defendant for damages in excess of Ten Thousand Dollars.

Respectfully submitted,

Diane Gore
Attorney for Plaintiff
OTTO INDUSTRIES

"What a bunch of legalese," Skaff thought. "In Assumpsit!— a fancy word for violating the contract; why can't lawyers use common expressions?" Appended to the complaint was a listing of the damages:

DAMAGES

I. OI cost of installing alternative piping system:	$1,000,000
II. Lost production for downtime:	$2,500,000
Total Known Monetary Damages:	$3,500,000

"Three and a half million dollars damages on a job worth three million, and the profit was $500,000. That's quite a bill," thought Skaff. "It's time to fight fire with fire." Skaff called his lawyer, Bill Pankratz.

DEFENDANT'S ANSWER

Count 1

1. Skaff admits that OI entered into a contract with Skaff.

2. The performance warrantees were specified by OI; proof of this averment [formal assertion as fact] is demanded.

3. Paragraph 3 of the complaint is specifically denied. To the contrary, OI undertook responsibility to assure proper interface and coordination between the system and the boiler.

4. Paragraph 4 is denied; Skaff avers that OI failed to operate the system according to Skaff instructions.

5. Paragraph 5 is denied as stated. Skaff admits that scaling occurred. Skaff denies the allegation that the scaling was extensive. Such allegation is a subjective conclusion drawn by OI which precludes Skaff from being able to form a belief as to the truth of this averment.

6. Paragraph 6 is denied as stated. Skaff admits it recommended and OI agreed to add seed slurry to the wet scrubber. Skaff is without knowledge or information as to whether OI correctly implemented Skaff's recommendation to add seed slurry because the means of proof are within OI's exclusive control.

7. Paragraph 7 is specifically denied. None of the alleged mechanical and hydraulic difficulties are the result of design deficiencies of the system.

8. Paragraph 8 is specifically denied. Skaff's system is satisfactory as designed. Moreover, OI has prevented Skaff from undertaking additional work or repairs upon the system.

NEW MATTER

First Affirmative Defense

9. Skaff fully satisfied its contractual duties by designing and installing a wet scrubber system, and has not breached its contract.

Second Affirmative Defense

10. The addition of seed slurry recommended by Skaff will prevent adherent scale. The contract does not preclude or restrict the use of seed slurry. OI has repeatedly added seed slurry to the System and has waived any objection to the use of seed slurry.

Third Affirmative Defense

11. The scaling was caused by OI operating personnel not keeping sufficient Dynamo chemical in the operating system.

Counterclaim

12. OI has refused to pay Skaff $150,000 under the contract even though Skaff, through its attorney, has repeatedly demanded payment. OI has breached its contract and owes Skaff $150,000.

WHEREFORE, Defendant Skaff prays that this court enter judgment for $150,000 with interest in favor of the Defendants and against the Plaintiffs, OI.

Respectfully submitted,

_____

Bill Pankratz
Attorney for the Defendant

Pankratz explained to Skaff, "We need to deny everything at this point until we can sort out the details and see who did what to whom. The next step is to retain an independent consultant to review the records."

The battle shifted to the Experts. At this time neither side had a clear understanding of what happened at OI. Something had gone wrong. Scaling had occurred in the wet scrubber. Action had been taken to resolve the problem. No one knew the root cause of the scaling. OI pointed the finger of blame at Skaff. Dynamo had not worked. Skaff suspected that the OI operators had failed to add Dynamo in sufficient concentrations at crucial times. The testing at Skaff's lab clearly showed Dynamo was effective in preventing scale in the OI scrubber.

Burger at OI was in contact with his attorney, Gore; they were considering names of possible experts to retain. They came up with three choices: Dr. Bill Morris, from the State University; Mr. Keiner, a consultant; and Mr. Matthews, a retired researcher from OI's parent company. Gore quickly got on the phone and arranged interviews with Morris and Keiner.

Gore eliminated Matthews because of possible jury bias. Gore was concerned that Matthews' past connection with Otto Industries would prevent him from appearing objective. Keiner, although impressive, expressed potential ethical problems testifying against fellow engineers. In the interview with Morris, Gore realized that the professor, although he had all the right credentials, was mainly a theorist with not much practical experience. However, with Keiner's reluctance to testify, Morris was retained.

Skaff's attorney, Pankratz, quickly retained a mechanical expert, Dr. Albert Lowry, whom he had used previously. Pankratz did not interview anyone else; he liked working with Lowry. Skaff had some doubts about Lowry's ability to understand the chemistry, but went along with the selection.

Over the next six months, Dr. Morris visited the OI facility, talked with the operators and engineers, read the available documents, and started working on a mathematical model of the wet scrubber chemistry. The story he pieced together was startling. Dr. Morris discovered that the source of coal to Otto Industries had changed just after their letting of the contract to Skaff. The new coal contained much higher concentrations of sulfur. More sulfur meant more scale; the greater scale overwhelmed the Dynamo chemical. In fact, it was impossible for Dynamo to work under these circumstances. OI engineers were not chemists. They had been unaware that their new coal source had altered the chemistry of the flue gas.

Skaff's Expert, Lowry, thoroughly investigated the mechanical and hydraulic problems. He concluded that OI had overreacted in changing out the lines and pumps in the system. The job could have been done for $300,000. Much of the damage sustained was due to the ineptness of the OI operators. With his limited knowledge of chemistry Lowry had no way of evaluating the effectiveness of the Dynamo chemical. But chemical evaluation seemed to Lowry to be unnecessary. Even if Dynamo didn't work, the maximum liability for Skaff should have been the $300,000 minus the $150,000 retainage, or $150,000. The $2.5 million liability for downtime appeared to be an inflated and unrealistic number selected purely for negotiating purposes. With regard to OI's operators, Lowry suspected they would have that much downtime during normal operational periods.

Fact Witness Depositions

Months passed with seemingly no progress on the lawsuit. There were no negotiations. Interrogatories went back and forth in a legal paper blitzkrieg. Then began a series of depositions. The opposing lawyers agreed to first depose the fact witnesses, beginning with Skaff and Burger.

OI's attorney, Gore, in conjunction with their Expert, Morris, decided to push Skaff on the issue of his responsibility to investigate the operations of OI, rather than relying on historical OI data as the sole basis for his design. Since the discovery period had not yet closed, information about the coal switch problem had not yet been supplied. Before Skaff would know what was happening, he would admit under oath that he had not conducted any study of OI operations. Then Gore would challenge Skaff with the theory that, had Skaff Engineers done their job and investigated Otto Industries' operations, they would have discovered that the coal source had been changed. Skaff Engineering should then have followed up with tests of their own.

On the other side, the Skaff Engineering attorney, Pankratz, wanted his client to emphasize the operator and engineering interface problems experienced by Skaff and his staff. Pankratz discussed with Skaff how to embellish his answers so that Otto Industries would be revealed as overly bureaucratic, incompetent and obstructionist.

Skaff's deposition revealed a lot about his own operations, and gave new insight into the technical problems of the case. Skaff was asked in detail about other projects using Dynamo. Skaff admitted that at one previous job of Skaff Engineering, scaling had been exhibited in a pilot-scale test at a plant similar to OI. The project was stopped. Skaff had theorized in that case that the coal was high in calcium. Skaff had not reported these test results to OI.

Under questioning:

Q: Why didn't you report the pilot test results you had run the previous year?

A: It never crossed my mind. We ran many pilot tests— some worked, some didn't.

Q: Why did you not run pilot tests at OI?

A: The time schedule was too tight, and we were confident the OI tests were okay.

Q: Do you still think the OI tests were okay?

A: With their lousy staff, I'm not sure anything was run right.

Q: If you had to do it again, would you pilot?

A: Yes, if I were given more time.

Q: Were you told by OI you could not pilot?

A: No, but they knew there was no time.

And on the issue of operator competency:

Q: Did you ever tell Burger of OI that his operators were unsatisfactory during the training course?

A: No.

Q: If the operators did not comprehend the material, is it not your responsibility to inform OI management?

A: They knew what kind of operators they had.

After six hours, the deposition was over. Both sides felt they got what they wanted.

Burger was next. Pankratz's strategy was to get out of Burger the fact that he had competent engineers on his staff to interact with Skaff's people. Therefore, OI shared in the responsibilities to make the system work. Also, he wanted to get on the record that OI had done a thorough investigation of Skaff and found him to have an excellent reputation.

Under examination, Burger readily admitted that Skaff Engineering had a good reputation, but denied that his engineers had any more responsibility than to check Skaff's drawings for errors. One sidelight issue came out:

Q: Were you solely responsible for making the decision to remove Skaff from the jobsite?

A: No.

Q: Who else was involved?

A: The Company President, Mr. Stanley Otto.

Q: What did he tell you?

A: He wanted to give them more time, but I was exasperated and convinced him to concur in my decision.

Depositions were taken of the other fact witnesses for both sides. Each side was building their case. Next came the Experts.

Dr. Morris had been working on his Expert Report and was ready with a draft. Gore was not happy with it. The report was not the adversarial document she wanted. Morris had dryly presented the facts with his conclusions—well documented to be sure, but devoid of accusatory words as to who did what to whom. "What do you expect from a professor?" Gore thought. "Before Dr. Morris is deposed by the other side, I will have to work with him, raise his energy level, so the Doctor will be able to prevent the purity of his technical conclusions from being muddied."

Skaff's attorney, Pankratz, had no such problem with Lowry. His Expert Report strongly supported the view that OI had handicapped Skaff, then wasted money in solving the problem they had created. Sometimes Lowry liked to pretend he was an attorney and role play with Pankratz. Pankratz liked that. At times they would pretend they were the opposition and theorize what the other case would be. Lowry had a forceful manner, which made him a tough witness for the other side to depose. At trial, Pankratz would need him to tone down his demeanor to command respect from the jury. That time would come. For now, Pankratz liked Lowry's pizazz.

Expert Reports

At this point Morris and Lowry were divulged as Experts in the proceedings and their reports were formally submitted.

EVALUATION OF THE FAILURE
OF THE SKAFF WET SCRUBBER SYSTEM
Expert Report by Dr. Bill Morris, Engineering Consultant

Summary

The Skaff Water Treatment System was designed to remove dust and other contaminants from wet scrubber water from the boiler at Otto Industries. The scrubbers designed by Skaff were to be used to remove particulates from the gas exhaust in the boiler operation so that air pollution standards set by the Environmental Protection Agency could be met. The main technical problem was the propensity of the contaminants picked up in the water spray to form scale deposits in the scrubbers, thereby plugging the equipment and causing failure. The Skaff System was to be designed to specifically prevent such a catastrophe.

The Skaff System failed to perform shortly after startup. Within three weeks, scaling in the scrubbers shut down the system. The reasons for the failure are divided into four categories:

1. *The wrong type of system was selected.* The Skaff System could never have prevented scaling. The design was based on the erroneous concept that the addition of an inhibitor chemical known as Dynamo would prevent scaling in the scrubber. They should have devised a system that circulated seed crystals in the scrubber so that any scaling would be on the seed and not the equipment. They did not incorporate that conventional method in widespread use by the industry in their design.

2. *Skaff did not test their untried system in this new application.* The boiler plant was different from other applications with which Skaff was familiar. They failed to recognize the differences, and, therefore, did not do what is conventionally done, i.e., test a small-scale (pilot) unit to see if it would work. If they had, they would have found that the proposed design would not have worked.

3. *The technical basis set out by Skaff for design was inadequate.* Skaff did not properly collect, analyze, or use the data they collected. The composition of coal changed after the initiation of the project. Skaff did not check to see if this change would have an impact on the system. This was evidence of a fundamental lack of understanding on Skaff's part of what was required for the project to be successful.

4. *After the system failed, Skaff made recommendations that would have further doomed the project.* The suggested fixups were unrealistic and expensive. Their ideas were untested Band-Aids on a system that had already failed. They never recognized, or understood the reasons why—that sulfate pickup was much higher than design.

Skaff had full design responsibility for the project through the "turnkey" contract with Otto. They failed in their responsibility to provide a workable, economic water treatment system that would keep the scrubbers free of scale. Otto made the correct decision to abandon the system and find another way to resolve their water problem.

Narrative

0.1 Introduction

In the latter part of the 1970s, the Environmental Protection Agency (EPA) proposed tightening the regulations on smoke emissions from boilers. They set deadlines for attainment of standards for every plant, including the boiler plant at Otto Industries.

In the boiler process, hot air picks up fine particles and gases. It is the particles that must be removed so that the gas can be discharged into the atmosphere.

Scrubbers are conventionally used to remove particles from the gas. With wet scrubbers, water is sprayed into a chamber to contact the particles. The particles are trapped in the water, which falls to the bottom of the chamber. The water now contains particles, lime dust, and dissolved gases, such as sulfur dioxide. These materials must be removed so that the water can be reused in the scrubber system. Otherwise, the lime will dissolve and form calcium, which will react with sulfate formed from the adsorption of sulfur dioxide. The two molecules, calcium and sulfate, can react to form a hard adherent scale. When scale forms in the scrubber, it cannot process the gases readily because the pressure builds up in the boiler system. Scaling in the scrubber is the most critical factor in the operation of the system. If scaling occurs, the system fails.

Two approaches are generally used by engineers to prevent scaling in the scrubber. They are used in tandem. A fraction of the scrubber water is treated to remove the particles and the scale-forming materials, primarily calcium and sulfate. The water is chemically and physically treated in a separate treatment system and returned to the scrubber. Furthermore, in the scrubber itself, particles are added in the water as seed so that if scaling should occur it will happen on the seed particles and not in the scrubber. This is conventional flue gas treatment technology for wet scrubbers.

0.2 Design

The predicted pickup rate for sulfate was greatly underestimated because Skaff failed to take into consideration that the new coal had a higher sulfur content. This hurt the design in terms of equipment sizes and the ability of an inhibitor to prevent scaling. The equipment would prove to be greatly underdesigned and capable of handling only about one-third of its design loading.

The more significant error was in the omission of a seed system for the scrubber. The final design had water with scaling tendencies, without seed circulating in the scrubbers. The predictable result was scale formation on the surfaces in the interior of the scrubber.

The greatest mistake made by Skaff was the omission of seed particles in the scrubber for the design. Without seed, the supersaturated water will scale the equipment. Even if Skaff had designed everything else perfectly, this omission alone doomed the design to failure.

0.3 Operation

Startup was commenced. However, within three weeks scale had plugged the demister pads in the scrubbers, creating high back pressures and forcing shutdown. The scale was determined to be calcium sulfate. The speed at which the scale formed was an indication of very serious problems. A plethora of mechanical problems were plaguing operations at the same time.

0.4 Shutdown

Otto found another way to deal with the scaling problem, namely to introduce seed into the wet scrubber. The Skaff system was shut down when the alternative proved feasible.

Otto had relied on Skaff and its expertise throughout the course of the project. Skaff did not meet reasonable engineering expectations because it made significant errors and omissions that led directly to the failure of the system. Table 0 illustrates these points. It shows the various pickup rates for calcium and sulfate. Note the unrealistically low numbers for sulfate selected by Skaff as a basis for design. Skaff did not recognize the inherent problems in the design numbers, nor the fact that the plant they designed produced water that adsorbed much higher amounts of sulfur dioxide than predicted because of the high pH. Otto took the proper action of finding a feasible alternative and abandoning the Skaff system.

Table 0. Pickup Rates of Calcium and Sulfate[a]

	Original Design Proposal[b]	Data Calc.	Data Calc.[c]	Literature[d]	Design Specs	Actual[e]
Reference:	Skaff	Skaff	Morris	Morris	Skaff	Morris
Calcium	1.8	1.46	1.82	1.53 (3.3)	2.2	1.1
Sulfate	6.6	3.08	4.84	5.77 (8.7)	2.76	10.5

[a]All data reported as grams per 1000 standard cubic feet of gas (g/1000 scf).
[b]Data provided by Otto.
[c]Numbers calculated by Skaff.
[d]Average reported readings, highest reported readings in parentheses.
[e]From computer model and actual data.

OI vs. SKAFF
EXPERT REPORT
by Dr. Albert Lowry

My opinions can be summarized as follows:

1. OI did not divulge all the data and information necessary for Skaff to complete the work.

2. OI did not cooperate with Skaff in the design, construction, and startup of the wet scrubber system.

3. OI did not provide competent operators for the new system.

4. Skaff had insufficient time to remedy the startup problems, which were largely mechanical in nature and are common when equipment is commissioned.

5. OI abandoned the system during startup because they found a less expensive method of operation.

Sincerely

Dr. Albert Lowry

Skaff was outraged by Morris's technical theory that the scaling was caused by the change in coal composition. Why hadn't OI told Skaff about that? The theory made sense. Skaff ordered his chemists to check the theory in the laboratory.

"No substantive information in the exceedingly biased Expert Report from Lowry," remarked Gore. "It is a typical smokescreen defense blaming the company for all the errors." The case was shaping up as another example of strong arguments from both sides, with the jury deciding whose perception of reality was the so-called "truth."

Maybe the time was ripe for settlement negotiations. Gore called Pankratz and set up a meeting with Skaff and Burger. Before the meeting Gore asked Burger what the bottom line was. Burger responded that he wanted the million dollars for redoing the system. When a similar question was put to Skaff by Pankratz, he said he would pay $350,000 just to get out of the lawsuit. He would split the million in half and subtract the $150,000 retainage owed.

At the meeting Gore threw out the figure of two million, and Pankratz countered with an offer of $100,000. Then neither side budged. But communications lines were now open. After the meeting, both lawyers talked in private. Both offers were flexible, but too far apart for settlement at this time. The Experts would be deposed next, heating up the war.

Expert Depositions

As a general rule, during a deposition an Expert Witness can only lose. The best possible technical case has been laid out in the Expert Report. At the deposition, the opponent's challenge is to knock holes in it and discredit it using the author's own voice. Equivocal statements are blown up to incredible proportions during examination. Lawyers tell their Experts to divulge as little as possible other than what has been written, particularly in situations like this, where there are no ongoing settlement talks.

Dr. Morris was examined first. The weaknesses of his report were quickly exposed:

Q: Did you run any experiments to verify the chemical explanation of scaling?

A: No.

Q: So the results are strictly from your computer model, correct?

A: Yes.

Q: Has this computer model been tested under conditions similar to the OI situation?

A: Not exactly.

Q: What do you mean?

A: The model was calibrated using the data collected during startup.

Q: Doctor, would it be fair to say that the model results are only as good as the data?

A: Yes, that's true.

Q: So if the data is bad, the results could be garbage.

A: Could be.

Q: So, "garbage in, garbage out"?

A: It's possible, but I think the data was good.

Q: Doctor, did you collect the data yourself?

A: No.

Another area of vulnerability was exposed:

Q: Doctor, have you ever designed or operated a wet scrubber system?

A: No.

Q: You are an Expert in chemistry, isn't that true?

A: Yes.

Q: You are not a wet scrubber Expert, are you?

A: No.

So, Dr. Morris has stated for the record that his computer model results were solely based on data collected by OI personnel during startup, and that he was not an Expert on wet scrubbers. The Defendant's line of attack was now clear.

The Defendant's Expert, Lowry, had his own problems. Under deposition examination he admitted:

Q: Are you an Expert in chemistry?

A: I have some practical knowledge of chemistry.

Q: You have no degree in chemistry, do you?

A: No.

Q: You do not perform chemical tests, do you?

A: No.

Q: You have never been qualified in a court of law as an Expert in chemistry, have you?

A: No.

And on the issues of the replacement of pumps and lines by OI:

Q: You did not perform any inspections of the pumps and lines that were replaced, did you?

A: No.

Q: You would design the system differently if you knew seed slurry was to be used?

A: I would change some things.

Q: What would you change?

A: I would put in more abrasion-resistant materials.

Q: So it's fair to say that material replacement that makes the system more abrasion-resistant would be required?

A: Yes.

Skaff's Expert has disqualified himself from dealing with the chemistry issues and has admitted that material replacement would be necessary to use seed slurry.

Both sides gained considerable knowledge through this portion of discovery.

The trial date was set. In sixty days the action would shift to the courtroom. In the meantime, both sides would be in top gear, preparing for trial: a final list of witnesses had to be selected, and each witness had to be prepared in detail. Drafts of examination questions were necessary. And all the documentary evidence and exhibits had to be complete and totally organized. Maybe there still was a chance to settle the case. Phone calls quickly extinguished that notion, however; both sides were frozen in their positions.

THE TRIAL

Pretrial

Presenting the Plaintiff's case, Gore boiled down the issues into three, in which she would seek the jury's affirmation:

1. Was the scale warranty in the contract breached?

2. Did Skaff screw up the design?

3. Was OI correct in removing Skaff from the jobsite and redoing the system?

Likely, Skaff would reluctantly admit the scale warranty was not met. Gore conjectured that the counter argument would be that OI failed to tell Skaff of the change in coal composition. In response, Gore would hit hard on the point that Skaff represented himself as the Expert in the field and OI relied on that expertise by the issuance of a turnkey contract. Skaff knew about coal composition and its effect on wet scrubber operation. Skaff's engineers should have asked OI about it. OI did not know enough to realize the connection or they would have relayed the information.

Pankratz, on the other hand, planned to emphasize the bureaucratic nature of OI's organization and how difficult it was to get responses and information. Furthermore, the relevance of the coal composition was so obvious that any OI engineer would have known its impact on the wet scrubber. Pankratz would show OI's culpability in their secretiveness and negligence in conveying valuable information to Skaff. OI did it to themselves. Skaff would vigorously attack the large expenditures for system modifications. If he couldn't win on the liability issue, he at least wanted to cut down the damages.

Gore decided on a lineup. First she would put the company lawyer on the stand to explain why a wet scrubber system was required to meet the EPA air pollution regulations. Then she would put on several operators to show what happened. Burger would follow, with Dr. Morris acting as the cleanup hitter for the technical side. Lastly, an accountant would confirm the damages claim.

Pankratz would rely solely on Skaff and Lowry, and possibly call Mr. Otto as an adverse witness if the Plaintiffs did not call him as their own witness.

The judge was a good one; intelligent and alert, he liked cases with technical complexities. A jury of twelve with two alternates was quickly selected. The trial was ready to begin.

Gore's style was to be a straight, matter-of-fact questioner. She relied on the jury to pick up the subtleties. Pankratz was of the old school. Loud and blustering, he liked to upset the emotions of the witnesses as he put them under fire. The contrast between the styles of the two attorneys would be interesting.

Opening Statements

First the Plaintiff's Attorney:

"Ladies and gentlemen of the jury, I am Diane Gore and I represent Otto Industries. I will be asking you to award the sum of three and a half million dollars to my client for damages sustained during the construction and operation of a wet scrubber system designed and constructed by Skaff Engineers. Skaff had represented to OI that they were Experts in this field, so OI relied on their representations.

"The wet scrubber is like a big washing machine. The waste gas from the boiler is mixed with water and the dust is removed. The big problem with scrubbers is scaling—the formation of a chemical deposit in the system. The gas has difficulty flowing when the scrubber scales up.

"We will show that Skaff warranted that the system they designed would not scale. Yet when the system was started up, it did scale and we had to shut down the plant. We will further show that the system as designed never worked, and that we had to modify it at great cost. The money we spent to change the system, plus the money lost while the system was down, are the damages we are asking for."

Now the Defense Attorney:

"Jurors, my name is Bill Pankratz, and I represent Skaff Engineers, the defendants in this lawsuit. Ms. Gore has portrayed the case as a simple one. Skaff promised to install a system that would not scale, but it did, and OI wants to be compensated for their losses. We admit that the system scaled. However, the reason it did was that Otto Industries did not tell us they changed coal suppliers, which altered the coal composition and the dust. The new dust scaled. We relied on data supplied by OI, and had a right to rely on it.

"Furthermore, we will show that OI spent much more money than necessary to make modifications to the system; plus, they still owe $150,000 to Skaff for the job. Lastly, the conversion to the seed slurry system has resulted in great savings to the operation of the system, which more than compensates for any losses they may have incurred.

"With the evidence presented, we are confident you will find that OI is entitled to no monetary damages and that we contractually need to be paid the $150,000 still owed."

As the trial progressed, through cross examination Pankratz was establishing that OI was saving considerable money by switching to the seed slurry system. Because OI did not have to pay for the expensive Dynamo chemical, the resultant savings could potentially offset the damages claimed by OI. This scenario put a big hole in OI's case. Gore realized there was no way to deal with the new issue at this time.

Witnesses

After the Otto Industries witnesses were called, it was time for the Expert, Dr. Morris. The battle between Pankratz and Morris was tough—the calm Doctor pitted against the emotional examiner. The Doctor got the upper hand in his crossfire:

Q: Dr. Morris, did you know that Skaff has designed several hundred wet scrubber systems?

A: That's unfortunate.

Q: Did you know all of them have been successful?

A: That's not true. I know of one that failed. This one!

Later, Dr. Morris had to agree with the chemical savings offset claim. However, Pankratz did not establish that chemical savings had been considered in the decision to switch.

After the accountant tap-danced on the damages, the Plaintiff, OI, rested. The Defendant countered with the *pro forma* request for a directed verdict, attempting to show that the plaintiff had made out a *prima facie* case. The judge denied the motion.

At this point in the trial, had OI proven with a "preponderance of evidence" (which, in legal terms, means more than 50 percent) that Skaff had breached the contract? The evidence was strong that scaling had occurred in opposition to the language of the contract and that OI had spent much money to redo the system.

The damages offset issue was the wild card. Pankratz felt confident at this point. The weaknesses in OI's case were exposed. He had to keep hammering away at the damages and offset issues, which were his strength. Skaff's laboratory had confirmed Dr. Morris's theory on the chemistry. He would leave that area alone. Lowry did not know chemistry, and none of his fact witnesses had evaluated the coal composition and scaling chemistry. He still had to make a decision on whether to call Mr. Otto as an adverse witness. It would help the defense because Otto would confirm that he had wanted to give Skaff more time. On the other hand, it was a gamble. Mr. Otto was a dignified, credible witness whose bearing would favorably impress the jury. No, he would not call Otto.

Skaff made a fine witness. He enthralled the jury with his rags-to-riches stories and his careful, methodical approach to engineering. On cross examination, however, he had to admit:

Q: Isn't it a fact that you did not consider coal composition in your engineering evaluation?

A: No, we relied on OI to tell us anything like that.

Q: Did you ever ask about the coal?

A: No.

Q: You knew, didn't you, that coal composition was important in wet scrubber design?

A: Yes, but we assumed the coal source was constant. The data showed no change over three years.

Q: You assumed that, didn't you? And your assumption was wrong, wasn't it?

A: Yes, it was.

Expert Witness Lowry's turn. After the perfunctory direct exam, Gore began her cross examination questioning:

Q: Mr. Lowry, you are not a chemist, are you?

A: I am knowledgeable about chemistry.

Q: But you are not a chemist by training or experience, are you?

A: I work with chemists.

Q: Your honor, will you instruct the witness to answer the question "yes" or "no"?

Judge: Mr. Lowry, please answer each question "yes" or "no," and then explain.

A: Yes, sir.

Q: Mr. Lowry, do you remember the deposition you gave?

A: Yes.

Q: Do you remember a question about your expertise as a chemist?

A: No, I don't specifically remember.

Q: Well, on Page 41, you responded to the following question:

Q: You have never been qualified in a court of law as an Expert in chemistry, have you?
A: No.

Do you recall that?

A: Yes, if that is what I said.

Lowry proved to be a tough, recalcitrant witness. His credibility with the jury suffered as a result.

The last witness for the defense was their accountant, who showed that, using Lowry's assumptions of actual damages plus deletion of downtime damages, the total liability was $150,000. Even this money evaporated when a $200,000 savings in Dynamo chemical purchases was shown. On cross examination, his testimony was not impeached.

Closing Statements

The jury was tired after two weeks of technical testimony. For the closing statements, the defense went first.

Pankratz:

"Jurors, you have patiently sat through two weeks of witnesses. At the onset I said I would prove to you that the fault for the scaling in the scrubber was OI's. Their own witness, Burger, admitted they knew about the coal switch and did not tell Skaff. Any technical person would know this fact was important. Thus, OI's own negligence was the main factor in the scaling.

"Furthermore, the damages claimed by OI were both inflated and offset by the savings in chemicals. OI is entitled to nothing. They caused the warrantee to be breached. Money awarded to them would be an unjust enrichment. On the other hand, we are entitled to the $150,000 still due on the contract. I pray you will find for us. Thank you."

Gore:

"If you had not been sitting here for the past two weeks and heard the testimony, you would find Skaff's argument very persuasive. However, the reality is much different. Skaff, not us, were the Experts in wet scrubbers. Coal composition is important. Skaff admitted that and knew it. Why didn't they just ask? We would have told them. They made it sound like we withheld information to cause the system to fail. Nothing could be further from the truth. We were under pressure from EPA. We had to have a system that worked. Skaff screwed up, then tried to blame it on us. They admitted scale existed and the contract warranty was breached.

"On the issue of damages. We wanted a first-class system. When it didn't work, we did not have the luxury of time to pinch pennies. We made the best decisions possible to expedite the changes and hold costs down. Remember, the fault was Skaff's, not ours.

"The savings in Dynamo chemical are real. But this is something we found out later. We did not make the decision to change based on savings. We were desperate. The fact that we have learned how to save money has no bearing on the facts of this case. The damages we claim are real. Say you own a car that is destroyed in an accident. If the replacement car costs less to operate, should those savings be offset from the money owed for the value of the destroyed car? Of course not. We are only asking for the cost of redoing the system plus the economic losses due to the downtime caused by scaling. We pray you will award us the $3.5 million damages owed."

After the eloquent appeals were completed, the Judge read the instructions to the jury.

Judge:

"First you must decide if Skaff did, indeed, breach the contract with OI.

"If you find that they did, you must consider the amount of monetary damages to be awarded. I have decided as a matter of law that the maximum damages that can be awarded are $3 million. Jurors, you may now begin deliberations."

The jury retreated to the Jury Room and a foreman was selected. They went around the room asking for opinions. Initially, two jurors sided with Skaff and four jurors with OI; six were undecided. They started arguing whether Skaff could pay if damages were awarded. Also, many of the jurors liked Gore and disliked Pankratz.

For the next several days, debate on the issues ensued. Ten of the twelve jurors agreed with OI, and that was sufficient for a verdict in OI's favor. In deciding on damages, they had to be in agreement. Because of the divergent opinions, they seemed to be deadlocked. Finally, in frustration, it was agreed for everyone to anonymously write down a dollar figure. An arithmetic average would be taken, and that would be the figure used. Those who were for Skaff could simply write in $0 damages. This was quickly done, and the jurors reported to the judge that a verdict had been reached.

Everyone reassembled in the courtroom. The judge asked the foreman if a decision had been reached.

Foreman:	Yes, your honor.
Judge:	And what is your verdict?
Foreman:	We find for the Plaintiff, Otto Industries.
Judge:	And the damages?
Foreman:	The sum of _____! [You fill in the blank.]

EPILOGUE

Otto Industries had won. But the verdict could have gone the other way. What made the difference?

As in many cases, significant arguments can be presented by both sides. The members of the jury must weigh their own opinions. In reaching a verdict, each juror looks for justification of his or her own decision. It's human nature. Subjectively, the impact of personalities involved in the case becomes associated with their side's position. The technical points of sympathetic witnesses are sure to carry more weight than the testimony of people who talk down to the jury, or are dogmatic.

Upon questioning after the trial, the jury mentioned the following:

The Experts

The jury based their decisions about the most critical issues, those concerning the technical case, on the testimony of the Experts. The jury felt Lowry was competent, but one-dimensional. He appeared only prepared to talk about the problems caused by Otto Industries, and how much money the equipment change-out *should* have cost, rather than the root of the original problem. He seemed weak in his technical knowledge, as a result.

Dr. Morris, on the other hand, had a better grasp of the technical issues; his command of the chemistry was apparent. Even though he never designed a wet scrubber, he understood the problems. His explanations were clear, simple, and credible to the jury. His politeness and quiet confidence on the stand made him, in the jury's eyes, more credible.

Personalities of the Lawyers

Pankratz seemed to enjoy insulting and intimidating witnesses. In the examination of Dr. Morris, Pankratz came across as a bully.

Gore was straightforward. A little boring, but with impeccable integrity.

The Clients

Mr. Skaff was better liked than Mr. Burger. An entrepreneur, Skaff was very interesting. His testimony was believable, but he was unable to shake their impression that his eye was more on the calendar and the cashbox than the job. He had been overconfident in Dynamo, even though another job similar to OI had been stopped in the testing stage.

Mr. Burger, like his attorney, was a no-frills guy, a big company bureaucrat. But he knew his stuff, and the jury believed Burger would have told Skaff Engineering about the change in coal suppliers if Otto Industries had known how important that was.

Both sides went into the trial with a shot at winning. Each side had the opportunity to present its case. The verdict was based on a complex interplay of perceptions and personalities. The jury deliberation process magnified certain aspects of the case, and downplayed other parts. In the end, a winner emerged from the murky and mysterious process. Justice had been served.

9

A Gas of a Case

WITZIG VS THE COUNTY OF SUGARLAND

I got a desperation call from local county attorney Margaret Scott. She had a case, a terrible case on the face of it, and needed expert help in figuring out if the county was responsible. Two hours later she was in my office with the following facts. A construction worker, Mr. Witzig, reported in early at his shop. He went to the ladies room to sit on the commode and have a cigarette. As he lit up, the room exploded, and he was burned on some vital areas of his body. Gas had apparently built up in this little-used bathroom, and his match ignited the combustible mixture. The county landfill was 1000 yards away. The attorney for Mr. Witzig was claiming the gas migrated from the landfill, and the county was liable. Would I please take the case?

I had reservations. My specific expertise was not in the field of gas generation from sanitary landfills. Fifteen years earlier I had done some modest research in gas generation from landfills for recovery of the gas and its use as an energy source. Another professor and I punched some holes in an old landfill and measured methane gas evolution rates. The project proved to be unfeasible and was stopped. A short time later, I consulted for a city on a potential gas problem. They wanted to build a civic center on the site of our abandoned sawmill, and soil cores indicated decaying mounds of sawdust. Shallow wells were dug, and small but significant gas production was noted. I recommended they remove the sawdust, and they did. I understood the mechanisms of gas production and its flow properties, but I was not up to date on the field and certainly did not claim to be a top expert. It was within my general area of expertise. I could handle it, but did I want to?

The second reservation was being a defendant's expert. Mr. Witzig was seriously injured and required many skin grafts. Although insurance covered the medical costs, he was entitled to pain and suffering compensation, and if the gas came from the county landfill, they should pay. I vocalized my concerns to Ms. Scott, the attorney. Whatever I found was okay. She just wanted to know the truth. Given her candor and desperation, I accepted the assignment.

For the next three months, I immersed myself in the landfill gas literature. I found that the migration of landfill gas to nearby buildings with subsequent explosions injurious to people happened about once per decade in the U.S. Landfills under good climatic conditions (warm and moist) can produce plentiful gas and migrate great distances through underground sand layers. The plaintiff's theory of the county landfill being the source was a real possibility, since the site was an abandoned sand pit in an old river bend.

A tour of the landfill by air and then a tour of the building on foot revealed some interesting facts. The landfill was huge with a three-foot clay layer on the bottom and side. Clay cracks during dry weather, and gas can escape through the fissures. In the building, the toilet was found to have been recently installed, and a hole was drilled at that time through the concrete floor larger than the sewer pipe. Gas could have easily migrated up through the annulus opening between the pipe and the concrete hole. My client looked like he was in deep trouble.

Were there other possible sources of gas? The area was punctured with abandoned natural gas wells. The wells were supposed to be concreted in and capped. But many were not. Could the gas have come from an abandoned well in the immediate vicinity? Could gas from septic sewage back up in these systems and leak through seams in the sewer pipe into the bathroom? These leads were worth pursuing.

The gas-from-abandoned-wells theory bombed out. None were found on the old maps. The gas-from-septic-sewage theory was promising. The building was located in a low lying area where rains were known to back up sewage into bathrooms. Plus, the sewer pipes were bedded in clay, which caused cracking of the lines. I pursued this theory with vigor poring over sewer line designs, flooding conditions, and sewer gas evolution rates. One glitch in the theory was that sewer gas is smelly. No odor was ever noted by any of the witnesses.

In the recent literature, I found a mathematical model of landfill gas migration adaptable to the Sugarland situation. I calculated all the inputs, and Eureka, the model spit out the answer. This is what I learned. Gas could have migrated to the building 1000 yards away, but the concentration was below the explosive limit. I ran a sensitivity analysis plugging in the best and worst case assumptions. The model results indicated it was more probable than not, greater than 50 percent, that the gas contribution from the landfill did not solely cause the explosion. However, some scenarios predicted it did. I wrote up my findings and prepared for the deposition.

Deposition

The parties had been talking settlement and were only several hundred thousand dollars apart by the time of my deposition. But their positions had hardened, and the atmosphere was very unfriendly. The deposition would be no fun; all sorts of loose ends were hanging out, and I was sure the opposing attorney would exploit these weaknesses.

From the opening questions, the deposition was weird. My qualifications were entered into the record without any questions asked. No questions about my background ensued. Instead, the attorney showed me every article in the literature about landfill gas explosion and asked me questions about each situation. He never asked me any questions pertaining directly to the case! I was both frustrated and surprised. I had primed myself to discuss my expert report, and he never addressed it. After a day of largely irrelevant questions (at least from my perspective), it was over.

Why did the plaintiff's attorney conduct such a deposition? My attorney supplied the answers. First, he was posturing for his client who was in the room. Second, he was not technically oriented and was not prepared to deal with science. Third, he just wanted to see how I reacted under oath to get a reading on me for trial. The deposition provided the forum he wanted, not what I wanted. It was his ball game, his questions, and I had to persevere.

Escape

The case was deadlocked and headed for trial. Both sides had believable theories with some evidence to back up the claims. It would be a trial of theory versus theory with the jury deciding the outcome. The other side had an injured man. Would the jury disregard evidence and give a sympathetic verdict? We needed stronger proof to win the case. If I drilled holes at regular intervals from the building to the landfill, I could trace the methane gas and have proof. It was expensive, but we would know the truth. My attorney agreed.

The first two holes drilled and sampled for gas yielded the truth. The gas was from the landfill and at high enough concentration to cause an explosion. The case was quietly settled. My lesson in this case was clear: doggedly pursue the truth and let the lawyers figure out what to do with the results.

10

Swamp Gas and the Greenhouse Effect

BAHAMAS INC. VS OSMAN GEOTECHNICAL

Often knowledge developed in one case is translatable to another case. A new apartment complex just outside of Sarasota, Florida was ready for occupancy. That morning, the leasing manager smelled some gas in one of the apartments she showed a prospective client. The maintenance man was called in to check the gas range for leaks. He did the wrong thing and tried to light the pilot. A tremendous explosion destroyed the apartment building and killed him and another worker. Television news broadcasts on the scene showed a bombed out structure leveled to the ground. The families of the two deceased men sued the apartment corporation Bahamas Inc., the gas company, oven manufacturer, and construction firm. In a defense move, Bahamas Inc. cross-claimed the geotechnical engineering firm who had conducted soil surveys for foundation design, based on the theory that natural gas from the adjacent swamp was the cause of the explosion. Osman Geotechnical should have known about the soil gas and divulged the information to the owners.

Lawyers talk to each other, and experts talk to lawyers. This informal network resulted in a call to me from the lawyer representing Osman. He knew I had finished the Sugarland landfill case and that everything I had learned about gas generation would be available to him free of charge. My job was to figure out how credible or incredible the swamp gas theory was. Although one lawyer talked about the theory in the most derisive terms, after Sugarland I was sensitive and wanted to convert possibilities into probabilities with few loose ends.

I took a nice mid-winter flight to Florida to start the investigation. Sure enough, a dense swamp was less than 30 feet from the leveled apartment building, not an uncommon scene in Central Florida — lots of swamps, lots of apartments.

I was given boxes of discovery material including a soil methane study that showed traces of gas around the complex. Questions similar to the Sugarland case entered my consciousness. How much gas can a swamp generate and at what concentrations? I needed to model the swamp and predict the extent of methane penetration.

The literature turned out to be very sparse on swamp gas generation. Lake Mendota next to the University of Wisconsin, Madison Campus was studied for decades. But it was a lake in a temperate climate. The agriculture literature was not very helpful. Rice paddies were considered significant sources of methane gas, but the swamp was not a paddy. Methane gas was a contributor to the greenhouse effect — the warming of the planet Earth due to carbon dioxide and methane gas build up in the atmosphere. I went deeper into the "methane as a greenhouse gas" literature and found some arcane data. Termites were identified as significant sources of methane to the greenhouse gases. The estimated two billion cattle on earth burped large quantities of methane gas. Interesting intellectually, but so what? What about swamps and their contribution to methane gas? I kept on manipulating the keywords on the computer literature search. Finally a break. An article appeared on the Pripet marshes in Russia. Methane generation rates were reported. I got hold of the article and had it translated. The literature cited led me to an article on the Sudd marshes in the African Sudan. Contrasted to the Pripet marshes in western Russia, the Sudd was in a subtropical region. But wait a minute, a brilliant idea hit me. The Everglades are in South Florida. Why didn't I think of that before? After researching the rest of the world, I finally went after and found just what I needed: gas generation rates in the Everglades.

I exercised the computer model. This time even the worst case scenario showed no appreciable gas generation rates from the nearby swamp. I decided to have a little fun. The cattle gas generation paper had some references to human gas generation. That article had actual human studies of flatulence. (I was happy I never pursued that course of study in school, but glad somebody did.) I compared the gas accumulation from the swamp into the apartment building and assumed in the worst case all of it accumulated in the one apartment. That amount turned out to be equal to the amount generated by two adults living in the apartment. What a beautiful comparison. Osman looked like they were not at fault.

Deposition

The deposition was a three ringed circus. All the attorneys from all the parties to the lawsuit were present. They obviously had been doing a lot of depositions together by the amount of joking and kidding going on. I put on my serious face and waited for the questioning to begin.

The deposition proceeded in the conventional way. Toward the end, I was given the opening to hit the attorney with my comparison. In my most serious tone, I described human flatulence gas rates compared to swamp gas rates in the case. After a ten second pause, everyone in the room caught on, and belly laughs ensued. The next week the two parties settled that portion of the case. Too bad. I really wanted to try out my punch line on a live jury. The lesson learned was one of greatest value. A metaphor or analogy that explains the case in simple and insightful ways can make a decisive difference.

11

It's Criminal:
More than Money at Stake

ENVIRONMENTAL PROTECTION AGENCY VS CHEM TANK

I got a call from a headhunter. They call all the time asking about my graduate environmental engineers. This call was unusual. He was looking for me. He explained a certain party was suing another one, and he was authorized to find an expert very knowledgeable on the federal hazardous waste regulations. The party was flying into Houston to conduct interviews. Would I be interested?

"Who was this party?" I asked.

"I can't say, now," he replied "At the interview, you will find out."

He asked if I knew of other experts from Texas only, and I gave him the names of several other professors. Was I interested? I was more curious than interested. Why the hush, hush treatment? I had been interviewed for expert witness jobs on many occasions and did not mind that. I checked my schedule, and it was open. I asked if they would pay me a fee for the interview, and he agreed. I accepted. I usually asked for fees when travel was involved or the interview was time consuming.

I was directed to one of the classier hotels, a penthouse suite. Waiting for me were four lawyers: two from the Environmental

Crimes Division of the Justice Department in Washington, D.C., one from the Environmental Protection Agency (EPA) Regional Office in Dallas, and one from the Houston U.S. Attorney's Office — a high powered crew. They had a case in which a huge chemicals storage company had been caught red-handed dumping hazardous wastes into the storm water drainage canal. They had no permit to discharge these wastes, which were tank and rinse remnants. They were supposed to haul them off to a commercial disposer. An employee inside the plant had tipped off the EPA to the practice, which had been going on for several years. He was still working at the plant and continued his deep throat activities. That was the reason for the secrecy. Was I interested in being an expert for the EPA?

Yes, I was interested. I had never been involved in a criminal case. Prosecutions for environmental crimes had increased dramatically, and I wanted to get on the inside to learn. Furthermore, I taught a graduate course in hazardous waste dealing with the regulations, so this case hit in the strength of my expertise.

The bad news was they had only fifteen thousand dollars total budget to support the expert. For the first time, I entered into a negotiation of my hourly fee. In the past, my fee was a take it or leave it one. These government fellows wanted to cut it in half. I finally relented and against my better judgment, agreed to cut my fee by 33 percent; they agreed to provide technical research and support so my time could be optimized. Both agreements were big mistakes, as subsequent events were to prove in their unfolding.

The feds had a war room in the federal building in downtown Houston. Three technicals were assigned to research the case. Earlier in the year the U.S. marshall had obtained a warrant to search Chem Tank. The raiding party not only caught employees in the act of dumping chemicals, but seized all their files. The war room was filled with these documents. Under criminal indictment were the plant manager and technical supervisor. They were facing prison terms up to 20 years, fines exceeding one million dollars, and had to raise bail to get out of jail. The seriousness of this case was clear. These men's lives were at stake. The EPA had the discretion to prosecute given the alleged willful abuse of the regulations and the contamination of local waterways by the illegal discharge of highly poisonous chemicals. They wanted to make these men examples to others who might think similar thoughts and follow through.

The lawyer assigned to me was a raw rookie just out of law school, but he had a technical background in oceanography. He was sharp, enthusiastic, and ignorant. This was his first case. He would later prove to be the most competent of the Washington gang, and I was glad he was dealing with me. The technicals appeared to be highly competent; they had taken part in the raid on Chem Tank and were highly motivated to put the badmen in jail. As the case progressed, they became difficult to work with. They had their own theories, different from mine, and pursued those at some cost to my case.

My theory of the case began with the fact that Chem Tank was a foreign-owned subsidiary of a Third World owned company. The plant manager was an international employee, not a local citizen, and he operated the plant like it was in the Third World. Furthermore, economics were a strong driving force as the company "saved" roughly a half million dollars per year by dumping rather than proper disposal. My theory of ignorance and economics was in sharp contrast with the EPA technicals theory of banditry, that these people were outlaw Mafia types. The technicals searched for evidence of conspiracies while I needed facts about amounts of hazardous wastes and costs.

On a mundane level, there were difficulties in communication because members of the war room "team" were in fact stationed all around the country: the para-legal was based in Denver, one technical came from Dallas, the other from Houston, and the attorneys were both located in Washington, D.C. but in different agencies. Only on rare occasions was this group physically working together. Most of the time they worked independently on the case, without the benefit of interaction and shared insight.

Halfway through discovery I found out one of my colleagues and friends was the expert for Chem Tank. He had first been approached by the EPA to be their expert, but turned them down because of the insufficient money. I liked him as an opposing expert as he was nonemotional and matter of fact. We had opposed each other in several cases and had mutual respect. He had a country boy demeanor, which worked well on juries as opposed to my teacher, professor demeanor, which seemed to work equally as well.

The differences between a civil case and a criminal one are striking. The plaintiffs in criminal actions must prove their case "beyond a reasonable doubt" as opposed to the civil standard of "a preponderance of evidence." The criminal standard is much tougher, so the case must be much tighter to gain a favorable verdict. Furthermore, the jury must be unanimous in the verdict, instead of ten out of twelve needed in a civil suit. In the discovery phase, experts are not deposed, so the potential for surprise is much greater. One difference in the laws for environmental crimes compared to normal crimes was the plaintiffs did not have to prove the defendants had criminal intent (in legal parlance this is called *mens rea*), only that they performed a criminal act.

Criminal cases come to the courtroom quickly because of the speedy trial provisions in the law. The U.S. EPA vs Chem Tank jury selection in federal court was done in one day. The judge essentially picked the jury. Right away one of the Washington lawyers on our side picked a fight with the judge on a procedural matter and lost. From that point on, the animosity increased between the two. As a result of this, and his own philosophical sympathies, the judge was openly against the EPA. That proved to be a tough obstacle to prosecuting the case.

Our case was straightforward. We presented fact witnesses, mainly Chem Tank employees, who testified the company regularly dumped hazardous wastes into the storm sewers. But the presentation did not go well. Our star witness, the secret informant who continued to work at Chem Tank, froze on the stand after his identity was revealed. In court, facing his boss, when asked to verify incriminating practices of his plant manager our "deep throat" could barely remember his own name. Cross-examination reduced the value of his testimony even further. I was placed as the cleanup hitter, the last person to testify for the plaintiffs. I had an additional burden of introducing into evidence much of the company records. The morning before I was to testify, the judge told my attorneys I could be on the witness stand only one half day, not the three days as planned. The questions for direct exam had to be severely truncated.

The direct exam started badly. My young and enthusiastic attorney did not know how to properly phrase his questions, which were objected to and sustained by the judge. After 30 minutes, the judge stopped the proceeding and admonished my attorney complaining he (the judge) was not a law professor and the courtroom was not a classroom for teaching him how to ask questions. He sat down, and a more experienced, but ignorant, attorney proceeded with questions from the yellow pad.

Cross-examination was devastating to me and the case. The technicals who prepared the documents for me did not read and interpret the operator log books carefully enough. And because I relied on them, I did not study the log books sufficiently myself to understand the notations clearly and the nuances. I stuttered and stammered and tried to think on my feet. Finally the torture session was over; it felt as if I had left behind on the witness stand a quart of sweat and two pints of blood. I silently vowed never to allow others to prepare my case.

The case continued to go badly for the plaintiffs. The defense attorneys used legal technicalities dealing with the definitions of a hazardous waste and exemptions such as *de minimis* provisions. Small losses of hazardous wastes were not regulated, and the defense tried the small loss argument. At the end of the trial phase, the judge threw out all but 12 of the counts against Chem Tank (there were originally 150 counts).

Amazingly the jury came back with verdicts of guilty on all counts. They had apparently seen through all the plaintiff's attorney bungling plus my not so great performance. The judge, seeing the verdict, immediately overruled the jury and threw out all counts on the basis there was "insufficient proof to convict."

What an incredible disaster, everything that could have gone wrong, did.

Afterwards, in the war room, the finger pointing began. I had not been on the losing side in the courtroom before and had never experienced the ignominy of defeat, particularly in the way the defeat was exacted. The EPA had imploded and destroyed themselves. The defendants walked away free. Well, not exactly free, I found out later their legal fees ran three million dollars.

The judge's throwing out the jury verdict was appealed. Eighteen months later the appeals court ruled the judge had abused his power and reinstated the jury verdict. Defeat had turned into victory. Justice prevailed from our perspective. The company ended up paying over a million dollars in fines. No one went to jail, but the ensuing publicity had the desired deterrence impact.

For me, it was the lesson of a lifetime. I will always prepare my own case because the bottom line is clear. It is my reputation at stake in the witness chair, and I cannot rely on others for what I need to know. Remember the special financial arrangement? It took me more than a year to receive payment. And a few months later I received a call from an investigator representing the U.S. Investigator General's office who asked questions concerning justification of my receiving "large payments" from the EPA. No acts of kindness go unpunished.

12

Turning the Tables:
The Expert as a Fact Witness

HETTMANSPERGER VS MACEDONIAN MANUFACTURING

I was on sabbatical at the University of Michigan, a time of rest and reflection, away from my normal duties and away from expert witnessing. I needed a break; I had been involved in three cases all of which went to trial within the past year. Too much. Fortunately, at Michigan I would be hard to find, so I thought.

A call filtered through from an attorney in Houston. We knew each other, but had not worked together. He had a case involving the design and operation of a wastewater treatment system at an industrial plant Macedonian Manufacturing. His client on an adjacent property stored construction equipment. They, the plaintiffs, were claiming salt spray from the wastewater plant had corroded their earth movers and cranes. My name had turned up in the initial document production. As a matter of fact, he noted, I had designed the system.

Panic crossed my mind. Was I also being sued? No. Macedonian was the defendant. I called the plant manager at Macedonian and asked him what was going on. Indeed, they were being sued in what he thought was a baseless action. Hettmansperger had purchased his property after the wastewater plant was in operation and complained almost daily about the spray issue. The plant manager did not deny salt spray sometimes drifted over the site. He had sent a claims adjuster over, but the report back was the equipment was rusting due to poor maintenance not salt spray. The manager gave me the phone number of his attorney on the case.

The attorney had been trying to track me down without success and was very pleased when I called. Yes, the treatment plant I designed was the subject of the lawsuit. So would I be both a fact and an expert witness? I did not like the idea of being both, but I knew that I would be involved one way or the other. So why not?

The problem of being both a fact and expert witness had emerged in previous lawsuits. Witnesses for the other side had claimed to be both, and the judges generally allowed them to be. They did poor jobs. They were perceived by the juries as biased for the side they were representing, and their testimony was discounted. Furthermore, they tended to be emotional on cross-examination due partly to the strong stake they had in the outcome. I enjoyed coming into a lawsuit as an expert detached from the emotion of what happened. I had no other stake in the outcome other than to tell the truth as clearly and cleanly as possible. In this case, I had no such immunity. I was not only involved, the credibility of my design was the main issue.

The Facts

A decade earlier I had been a consultant to Gebert Plastics. They had me design a fairly unique wastewater treatment plant, one that combined cooling with biological treatment. The structure looked like a cooling tower with fans on top. Therein lay the problem. The fans carried tiny water mist droplets containing salt into the atmosphere. I modeled the trajectories of the droplets along with the physics of evaporation and dispersion. The results indicated no significant increase in salt content of the atmosphere outside the property limits.

A year after the system was designed, the plant was mothballed and put up for sale. The plant was sold to Macedonian who started it up with all new personnel. I was involved in the not very smooth start-up. Batches of chemicals were dumped into the wastewater raising the salt content. The spray landed on equipment and cars leaving tell-tale white spots.

The salt spray problem took two years of memos and problem solving to lick. All these documents were discovered. During this period, Hettmansperger bought the adjacent vacant property and stored his equipment. I viewed the case as one in which newly minted operators were learning how to make plastics by trial and error. There was some culpability, but could Hettmansperger prove damages?

They had one critical bit of evidence in their favor. The local pollution control agency had been called out to the plant frequently by Hettmansperger and had cited Macedonian Manufacturing with numerous nuisance violations for the salt spray. The case was a tough one from any objective standard, and I was having a hard time being objective.

Emotions

I often marveled at how the fact witnesses in this lawsuit could be so certain of their knowledge of facts based on their memories, yet have totally opposite views of what happened. Their minds conveniently forgot certain elements, while remembered others that fit better with their paradigms. They were for the most part honest people struggling to remember occurrences from years before. However, after years of recounting, their stories took on a self-justifying tone. Even documentary evidence to the contrary many times did not dissuade these fact witnesses from what they knew to be true.

But that's human nature. We will fondly recollect events long past in the most favorable light, forgetting parts of the stories that do not fit. Our emotional attachments to life events alter reality in subtle ways. Experienced attorneys understand the ways witnesses distort facts and know when to be gentle and when to be rough with witnesses. Besides, their fact witnesses go through distorted remembrances of past events.

There I was, observing myself becoming emotionally involved in the case and thinking of ways to defend my actions in the best possible light. I remembered the days and nights spent with the plant personnel struggling to run the wastewater system. I loved those people and felt the urge to protect them and their decisions against those who would discredit them. To my horror, I started acting like an advocate to my attorney trying to persuade him as to the righteousness of the decisions made. But he had the facts on paper in memos. Was I going to be able to control the monster inside, which was pushing me from unbiased expert to biased advocate?

I directed my thinking more on the dilemma of just the expert witness. Biased experts look and act like hired guns willing to say whatever the attorney who pays the bill wants them to say. Do we all become biased to a certain degree because we form relationships with only one side of a case and see the other side through the adversarial process? Yes, the tendency is always present. That is one reason experts on opposing sides can come up with such differing opinions.

Deposition

I was never so apprehensive at a deposition. The monster inside was trying to control me. The only thing I had going for me was the opposing attorney who for whatever reason treated me with respect. I did the best I could which was not very good. My attorney fully understood my dilemma. Three weeks later he settled the case and allowed me to go on with my life. My lesson was to deal with the forces within the human psyche that can distort facts and do my best to control and eliminate bias that could destroy me as a credible witness.

13

The Big One

THOMAS *ET AL.* VS NOBLE OIL

Every football player dreams of winning the Super Bowl. Experts dream of being in the big lawsuit, a potentially precedent-setting one with millions of dollars at stake. As with Super Bowl games, these cases represent the ultimate test of the expert's skills. The lawyers on both sides are the best and can recruit the best experts with the lure of plentiful resources. Prestige, honor, machismo, and credibility are all at stake in the Super Bowl of all lawsuits, known as the Big One.

The call came in from an attorney. He wanted to retain me as an expert in a hazardous waste site case. His name sounded familiar, Tom Parsons. My mind flashed back to three years earlier when he interviewed me for another hazardous waste case and did not follow up. Six months later I was subpoenaed to give a deposition in the case. Since I was not retained and knew nothing about the matter, I ignored it. Wrong decision. One cannot just ignore a subpoena. Anyway, I ended up mad at Mr. Parsons.

This time, Mr. Parsons offered me a ($10,000) retainer and apologized for past sins. He represented two thousand citizens who lived near an 11-acre abandoned sand pit filled with hydro-carbon-type hazardous wastes. It had been declared by the U. S.

Environmental Protection Agency as a Superfund Site, which meant it was really bad. A potentially responsible party (the ones who dumped the chemicals there in the early 1970s) was identified as Noble Oil Company, a large producer of petrochemicals. They were in the process of cleaning the site.

I told Mr. Parsons I'd think about his offer. My reluctance not only concerned his past bad behavior, but my experiences with two other similar cases. A decade before, I was an expert on a case where a disposal company had dumped toxic oily wastes on backwoods dirt roads making the homeowners very sick. I put eight months into the case when the attorneys ran out of money and had to accept a settlement of one cent on the dollar. No one but the defendants were happy with that outcome.

Two years later I was retained by the plaintiff's attorneys to investigate a Superfund Site adjacent to a middle class neighborhood. A medical doctor was also retained. As we progressed in our investigations, the doctor and I came up with very different potential causes for the observed health effects of the occupants. His analysis pointed to leaking of toxic chemicals into the drinking water even though samples analyzed indicated no contamination. My analysis concluded the pollutants came through the air route. The analytical results for air were equivocal since at times the upwind samples were more polluted than the downwind ones. The doctor dogmatically stuck to his position, and I doggedly stuck to mine. The opposing attorneys jumped on each of us with glee during the depositions and duly noted our vastly different interpretations. How wonderful, they must have thought, to have a case in which the opposing experts disagree as to the cause! The case settled at trial right before he and I were to testify. "The doctor should have stuck to medicine," I grumbled, "not the sciences of environmental contamination." But he thought he knew everything and was a danger to himself and the case. I had never been more uncomfortable in a case.

I worked on two other hazardous waste sites. One settled right after I prepared a technical report for the settlement brochure. The other one involved excruciating amounts of data subject to many interpretations, but it, too, settled. Experience told me waste sites were difficult. But they were difficult for all sides. Maybe this time I knew how to do one right.

I deposited the check and went to work.

Tom Parsons had just filed the complaint and was preparing the interrogatories and a request for document production. I grabbed several boxes of documents to familiarize myself with the background and to help him formulate questions. Then we waited several months for the requested documents to arrive. And arrive they did.

The approach was the "needle in the haystack" in which the opponents supplied tons of paper so overwhelming that my eyes became bloodshot reading only box labels. I spent all my time searching for meaningful documents in the flood of paper.

This was a big money case. One of my first jobs was to figure out what other experts to retain. I saw a need for many: a hydrologist to look at the seepage route; an emissions chemist to predict the volatilization of chemicals from the pit over time; an air pollution modeller to predict the plume of toxic chemicals and at what concentration these chemicals interfaced with the people living close by; an immunologist to discern patterns of health effects; a toxicologist and/or medical doctor to investigate disease; a real estate appraiser for land values; a demographics expert to examine any long term effects on the area; and an economist to evaluate the impact of the waste site on other amenities. Also, a standard-of-care expert was needed to reconstruct the waste disposal practices for the time period of the dumping. It was a tall order to find highly competent experts who had the time, interest, and willingness to be involved. So many areas of expertise were involved that careful attention had to be paid to make sure they each had the same information to work from (discovery documents, literature search information, etc.) and were kept informed of the others' findings to ensure continuity. Orchestrating such a large group of independent experts was a big role, but a necessary one in this case. As the lead expert, I was asked to assume this role.

The attorneys knew some experts in the desired areas of expertise, and I knew others. Some experts had conflicts, like one who had already been hired by the other side. She was a colleague of mine; I knew her well, and I knew she was sharp. I had the experts submit their resumes to make sure their qualifications matched expectations. The attorneys, of course, had the final say. The experts were brought on board.

In these complex cases, the work of one expert becomes the raw material for the next one. For the hazardous waste site, I needed to figure out how much and what kind of wastes were dumped into the pit over time. The chemist calculates the emissions, which go into the computer model of the air pollution expert. His output is used by the toxicologist/medical doctor to render an opinion on damage to the human body. The real estate expert looks for declines in real estate values as a function of distance from the waste site and compares these properties to other similar neighborhoods away from the site.

The process looks cut and dried, but it isn't. First of all, the reconstruction of the amounts and types of toxic chemicals in the pit was nearly impossible. No analyses of wastes in the pit were made during the early 1970s. Recent analyses represent only the fraction remaining after weathering and several floods, which carried away large quantities of sludges in the 1970s and 1980s. However, some finger prints of chemicals remained. Fortunately, invoices were retained by the trucking company who hauled the wastes to the pit. The invoices were vague in terms of types of wastes and referred only to the generation processes, such as "styrene tars" or "156F Bottoms." This sketchy information forced me to understand the chemical processes of the 1970s and figure out the waste products. Later on one of my assistants found a book with the information on processes and wastes generated. It was too late to save hundreds of hours of time already spent, but I was well prepared to deal with any process questions on the witness stand.

We experts got together and discussed theories of the case and who would testify to what. The theory of the case hinged on figuring out why the major players for Noble Oil dumped the wastes in pits. Environmental regulations were being implemented at the time, and the site was under fire. A court order in

1974 finally shut down the operation. From our conversation, it was obvious we needed one more expert, one familiar with the regulations at that time.

For openers, the question was did the Noble Oil managers know or should they have known that dumping in the pit was not a reasonable or accepted practice during that time period? Some companies were using unlined pits for waste disposal in the early 1970s, and the other side would try to show this was standard practice. On the other hand, articles in the literature during that period clearly indicated the problems and dangers of the practice. Internal memos by Noble Oil engineers indicated they were questioning the practice and looking for alternatives. But they did nothing and continued to send the wastes to the pit. The "standard of care" issue needed more work to sharpen the focus. What were the alternative modes of disposal? Could Noble Oil have built an incinerator or cleaned up and recycled the residues?

The "causation" issue was more difficult. The easy part was showing that Noble Oil deposited wastes from their processes into the pit. The tougher part was figuring out the pathways for escape of the toxic materials. Computer models were used to predict the emission rates and concentrations over time and space. Medical reports were needed to evaluate harm to the nearby residents. Lots of work.

Deposition Phase

Both sides started deposing fact witnesses. Our side started with Noble Oil Plant Manager Phil Xia. I sat next to Tom Parsons as the interrogation began.

Parsons: Mr. Xia, please tell the jury your academic background.
Xia: I received a BS degree in Chemical Engineering and Chemistry from Tulsa University in 1960. That was the extent of my formal training.

The response startled me. I had in my possession a paper he wrote on refinery management published in *Hydrocarbon Processing* stating he had a BS and MS in Chemical Engineering. Something was wrong. At the break, I showed Parsons the paper, and I flashed a wry smile. The interrogation continued:

Parsons: Mr. Xia, did you write this paper?
Xia: Yes.
Parsons: The by-line indicates you have a BS and an MS degree in Chemical Engineering. Do you have an MS degree?
Xia: No. It must have been a mistake or misprint.
Parsons: Did you check over the paper before submitting it?
Xia: I don't remember.
Parsons: Do you have a resume?
Xia: No.

That was a strange response. Everyone has a resume. He was hiding something. After the deposition, I had an assistant check with the University of Tulsa, and she found that Mr. Xia had a BS degree in Chemistry, not Chemical Engineering. He had lied twice, an interesting character flaw and one a jury would be attentive to. These deceptions were credibility killers on the witness stand, and Xia and his attorney knew it.

As the depositions of the fact witnesses proceeded, the strengths and weaknesses of both sides magnified. The defendants knew what they were doing when they deposited the wastes in the pit and knew toxic materials would escape. However, they had not been cited for any violations of regulations during that period, and other similar companies had done the same thing. The plaintiffs were having difficulties in showing actual damages in terms of lower real estate values and proximity problems specifically related to the toxic chemicals. Even though the standard-of-care and technical cases were strong, the truth was if damages could not be more precisely determined the lawsuit value would be low.

Settlement Negotiations

Contrary to popular belief, lawyers try to settle lawsuits right from the beginning. The rub is both sides want settlements most favorable to their clients creating winners and losers. Usually at the beginning of a major lawsuit, informed settlement discussions are begun. It can be as simple as the plaintiff's attorney over cocktails asking for ten million dollars to get rid of his case. He will preface his remarks with how terrible the adverse publicity will be to the defendant and that the case is really worth much more than that. The defense attorney will respond that she will discuss it with her client, but she thinks the case is only worth two million even though the plaintiff is asking for sixty million. The plaintiff's attorney goes away knowing that 40 percent (his share of the settlement) of two million is bankable and he has a case worth some merit. Since he is on contingency, the plaintiff's attorney gets nothing in cases with no merit. Fortunately, experts get paid regardless of the outcome. The defense attorney comes away knowing the name plate value of the case is greatly inflated, as it usually is. She also knows if the case goes through to trial millions of dollars could be spent in legal fees by the defendant. After a brief discussion with the client, the informal offer is not usually pursued. Instead, the plaintiff is asked to prepare a settlement brochure with details on the offer.

For Noble Oil, I was asked to put together the technical case as a chapter in the brochure. It was a recitation of the history of the site, the escape routes of the chemicals, and potential damages. With references, it was twenty pages long. The plaintiff's attorneys asked for thirty million dollars, and after due deliberation, the offer was rejected. The two million dollar counteroffer was restated with no movement by either side.

The most probable time for settlement is after discovery is complete and the depositions are over. At this point, more than 95 percent of the facts are known to both sides, plus the strengths and weaknesses of the case and the experts. If settlement does not happen before trial, it is usually because of the arrogance of the attorneys or belligerence of the clients. These emotions cloud the factual issues and block negotiations.

The Thomas vs Noble case continued. Halfway through the expert depositions, the sides agreed to mediate. This form of alternate dispute resolution was set up for a three-day period with a professional mediator. In attendance were the attorneys and clients for both sides. In this case, with two thousand plaintiffs, twenty individuals were chosen to represent the entire group; it would be their responsibility to get the agreement of the other plaintiffs. Mediation ground rules gave each side one-half day to present their case with no cross-examination or interference. On the second and subsequent days, both sides would be sequestered, and only the moderator would move between them to communicate. At the end of the first day, following the presentations, the mediator handed out their homework: both sides were to start the second day with their most reasonable offer. The next morning, the plaintiffs offered forty million dollars and the defendants six million dollars. Through an entire day of haggling, the plaintiffs came down to thirty-one million, and the defendants went up to nine million. A break came the next morning when the defendants offered fifteen million dollars, take it or leave it. The plaintiffs came back with twenty five million. The day wore on. Tom Parsons was willing to take fifteen million dollars, but his clients said no. Fearing that he would lose the deal, he convinced his clients to offer to split the difference and try for twenty million, take it or leave it. Noble took it, and the case was over.

A mixture of feelings went through me. Twenty million was a very fair settlement — a victory, a triumph of sorts. But I had fallen in love with the case. I knew more than anyone else about the technical details and had spent almost a full year organizing massive amounts of data into a cogent explanation, and it was now all obsolete information. All that hard work with no culmination left me empty. The lesson learned was not to get so involved in a case that I could not let go when it ended. Keep it all in perspective.

The next day I got a call from Tom Parson. He congratulated me on my fine work. Then he said, "By the way, I have two lawsuits pending. Would you be interested...?"

14

Junk Science

WATT ELECTRIC VS MG ASSOCIATES

Experts practice in fields of specialties, which are both broad and narrow. The breadth is established through some standard discipline like toxicology, metal fracture mechanisms, plant physiology, or water treatment with which the expert has established credentials such as formal education and work experience. Within the discipline, the expert establishes several narrow specialties. For instance, a toxicologist may have particular expertise on the effects of heavy metals on fish and particularly on lead in the aquatic food chain. Experts, particularly academic types, try to narrow their case specialties so they are the only, or one of a few, practitioners of some arcane quanta of knowledge. If the expert is lucky or charmed, at some point in his witnessing, a case will involve this narrow specialty, and the expert can claim to be one of the top people in the field.

It happened to me. I was an environmental engineer practicing in the field of industrial water and wastewater treatment. In the 1980s, most of the field was redefined by federal regulations to be hazardous wastes. I concentrated in a narrow specialty within the subject of industrial cooling water treatment. Industrial plants use a lot of water for cooling their equipment. The heated water is recirculated to cooling towers, which contain heat transfer media called "fill". The hot water trickles down the fill, and air is sucked up. The water loses heat primarily by evaporation to the air and is cooled. One of the big problems is that micro-organisms thrive in cooling water and can form biofilm on surfaces and destroy the heat transfer efficiency. That turned out to be the subject of the lawsuit.

Watt Electric sued the engineering firm who designed a new cooling water system for their coal-fired electric power generating station in Coalton, California. The unique part of the design was the use of sewage plant effluent as the cooling medium. This meant that a large source of microbes were entering the system at all times. The cooling tower fill was made of sheets of corrugated eggcrate-like plastic crates. The hot water trickled down and was cooled by the rising air. The biofouling was controlled by the addition of chlorine, a powerful biocide, into the cooling water. Within two months of startup, severe biofouling was discovered in the fill. Within six months, some of the fill was completely clogged. Expensive removal and cleaning of the fill was done. However, within six months of reinstallation, the fill plugged again. The basis of the lawsuit was negligent design. MG Associates knew (or should have known) that corrugated plastic fill would plug up if sewage plant effluent was used for cooling, according to the complaint. An alternative fill material called splash bars, which were plastic slots, could have been installed and would not have plugged up, claimed Watt Electric.

A call came in to the director of the Cooling Water Association from the lawyer representing MG Associates. She passed on my name to the attorney Denise Morris. The case hit squarely in my strongest specialty. I took it and soon thereafter was on an airplane to Coalton. I toured the site and discussed the situation with Ms. Morris. At first blush, my opinion was that the designers did, indeed, screw up. My feelings were mostly negative, but I indicated that I would see what could be found to mitigate what looked on the face of it to be an open-and-shut case. I was retained as a consulting expert, so that if my final opinions were not what the defendants were looking for, they had the choice not to use me as a testifying expert. At least I could give them the unvarnished truth, so they could decide how to proceed with the defense.

I asked Ms. Morris if she had any information as to who the expert for the other side was. She said yes, Paul Ogan. I knew him; he was an old colleague and friend, one of the few top people in my narrow specialty. We had done some consulting work together in the past. At one time, he was president of the Cooling Water Association, and I was the editor of the Journal. That made the case even tougher. Ogan was not only strong technically, but he had a magnetic personality and was a super salesman for his company. He had been a long-time consultant to Watt Electric on their cooling water problems.

Discovery

The plaintiff's law firm had a reputation for nastiness. They responded reluctantly to requests for documents and only complied when precision specificity was given. Slowly we made progress. Then we had a big breakthrough. I knew that operating companies such as Watt Electric normally bought water treatment chemicals from suppliers who also made calls and wrote up service reports at least several times a month. We got hold of these reports, and they were solid gold. The information showed the operators of the cooling towers never dosed the chlorine at the right intervals. When a chlorinator pump broke down, the maintenance people took months to repair it. The theory of our defense became very clear. The chlorination system was never operated properly. Inadequate chlorine allowed the fill to be clogged with biofilm. These service reports were made by a third party not involved in the litigation and were very credible. The service representative appeared to be impartial and competent. MG Associates were back in the ball game. They still had their problems. They could have used the nonplugging fill in the cooling tower design, and in my discussions with their design people, they did not evaluate plugging potential of the fill.

Lawsuits are never decided by lopsided scores. They are more like basketball games with one side winning 96 to 89 points. Both sides have strengths and weaknesses, which the lawyers exploit. At least we had uncovered a major strength in the third party reports.

Depositions

I had never before witnessed such antics in deposition. The strategy of the plaintiff's attorneys was to threaten, cajole, and intimidate the witnesses for our side and prevent our attorneys from questioning their witnesses. Nearly every question asked by our attorneys of their witnesses were objected to, and these people looked like they had been instructed to answer the questions as to not reveal anything. Strange and frustrating. What were they hiding?

Even Paul Ogan was evasive. He could hardly remember his name during the deposition. I knew he had a photographic memory, so his testimony didn't make sense. My deposition was a war. The attorneys spent most of the time arguing and threatening to certify questions and to go directly to the judge for decisions. Three days of this nonsense were enough, and both sides withdrew. This case was obviously headed for trial with lots of surprises ahead.

The Trial

Mr. Ogan was the cleanup hitter for the plaintiffs. During the discovery phase, we never talked to each other, even at professional meetings. This was not only a lawsuit — this was him against me. The whole small universe of cooling water professionals were observing the trial to see which gladiator came out on top.

In the direct case, Ogan developed a novel argument, so novel it went against all the conventional and accepted concepts. He argued that the surface area of the fill was the most important factor in chlorine dosing: the greater the surface area, the higher the dosage of chlorine necessary to maintain the surface of the fill free of micro-organisms. Since the corrugated fill had a much higher surface area per unit volume of cooling space, it required significantly more chlorine. He argued further that MG Associates never informed Watt Electric about the additional chlorine requirements and that the engineers sized the chlorine dosing equipment too small to add the order of magnitude for more chlorine that he calculated was necessary. The operators did not add sufficient chlorine because they were ignorant of the additional need and because of the inadequate dosing equipment.

He argued that even if the dosing equipment was beefed up the amount of chlorine needed was so great the operating costs would add millions of dollars to the overall price of the cooling towers making the corrugated fill decision by MG Associates a bad one.

Ogan's argument was a clever one. There was a common-sense appeal to his novel theory that fill surface area was more important in chlorine dosing. Furthermore, he was circumventing the main case against his client, the third party service reports, which clearly showed inadequate chlorine dosing, by blaming MG Associates for bad design of the chlorine dosing equipment. The problem was his theory was his alone — totally unsupported by the accumulated research and field experience of everyone else in the field. But here it was being presented for the first time, not at a technical meeting, but in a court of law.

The accepted theory has as its basis the population of living micro-organisms as the important and controlling factor. Chlorine is dosed to kill microbes down to below a certain level in the water. If the population is kept low, the microbes will not be in sufficient numbers to attach and grow on any surface. The amount of fill surface area is only important if the microbes are not controlled; in which case, the consequences can be devastating. The microbe concentration in the water is always the primary control variable independent of the kind of fill. Ogan's novel theory could only be classified as junk science.

In preparation for Ogan's cross examination and my direct examination, I had to do some soul searching. New and novel theories are usually rejected at first by the scientific community. The case of Galileo and accepted theories of the solar system flashed into my mind. He presented the heretical view that the earth revolved around the sun and was forced to recant. I could not portray Ogan or his theory as a crackpot even though it had no scientific credibility or evidence to back it up. On cross, we would have to get Ogan to admit the lack of basis for his theory and the weight of the literature supporting my theory. When my turn came I could forcefully argue the strength of my theory and the weakness of his.

Ogan completed his direct exam with a flourish. He and his attorney looked like they had choreographed the perfect duet, and with Ogan's extroverted personality, it looked like the jury was impressed. What did they know about science and proof? My duty would be to provide that education.

On cross-examination, Ogan got off to a very rough start. First, his demeanor changed radically. He had become nasty and abusive, unwilling to respond directly to the questioner. This line of questioning did him in.

Morris:	What was your consulting fee on this case?
Ogan:	My fee was between $120 and $150 per hour.
Morris:	Was it $120, $130, $140, or $150? Be more specific.
Ogan:	As I stated, my fee was between $120 and $150 per hour.
Morris:	You mean that when you submitted an invoice you stated, "Pay me between $120 and $150 per hour?"
Ogan:	Ah, mm, ah, mmm (stammer and stutter) I was paid between $120 and $150 per hour.

His credibility with the jurors faded, as not only was he not responsive to the question, but he was taken to the limits of absurdity and still did not respond. His change of demeanor put a chill on his credibility. The cross-examination on his theory was devastating. He had nothing to hang it on other than his experience (undocumented) and his opinion (unsupported).

After the plaintiffs' case was presented, they tried for settlement. The offer was so low, below the amount that our side had previously been willing to settle for, that we became confident of the final outcome. As a result, I went to the witness stand overconfident. On direct, I matter-of-factly showed through graphs, charts, and slides of the service reports my theory of what happened. I discussed Ogan's junk science theory and dismantled it piece by piece bringing together the contemporary scientific literature and melding with the design and operation of the cooling tower.

On cross-examination, I was not prepared for the nastiness to follow. Early on the opposing attorney attempted to put into evidence a rough draft of an article I had written on expert witnessing. My attorney objected, and it was not entered. However, it staggered me. The rough draft could only have been obtained from my files or close associates. Had they put a private detective on me? The nastiness continued. The line of questioning was that since I did not design cooling towers I was not qualified to give an expert opinion on the fill. I admitted I did not design towers, but that did not disqualify me from offering opinions on the chemistry of water treatment. At one point, I lost my temper at the ridiculousness of the argument and felt the insult personally. At the break, my attorney grabbed me in a forceful way, reminded me that I was the expert and did not need to mud wrestle. The jury would see through the smoke if I kept my cool. That was the only and last time under oath I let my emotions get to me.

The jury did come in with the verdict for our side rejecting the junk science arguments. I found out secondhand that Ogan was astounded his side lost the case. Our relations cooled to greetings from a distance. He took the defeat personally and hard. I did notice in his short-course brochures that he had an element on biocides using different cooling tower fills. People who took the course told me his beliefs in the junk science theory were genuine. He apparently had convinced himself of the correctness of the theory, and it was unshakable. It is amazing how his mind worked. But junk is junk, even in the courtroom with persuasive graphics and a witness with great personality and novel theories.

The bottom line was that the jury was able to make a proper judgment in this battle of theories. It took additional effort for me to explain to the jury, in terms they understood, the way science worked. Since then I have learned that I need to do teaching of the basics in every jury trial.

Section III

Observations and Conclusions

15

To the Lawyer

Hire an Expert consultant shortly after a client has contracted for your services. Hire the Expert for your education and for early preparation: a good Expert can not only give a good early reading of the technical issues but can also give insight into the potential value and variety of litigation. Legal and technical theories should be explored early.

As a preliminary step, hire an Expert to be your consultant on the technical issues of a case. I have been called upon by attorneys to be a generalist. My duties included the recommendation of Expert Witnesses in the litigation of hazardous waste sites, with the understanding that I would not be called as an Expert Witness myself.

Early interaction with the case is important to the Expert. As a technical detective, they can ferret out elements that can make a big difference later on. I was hired as an Expert for a big state university that was concerned about a large private development on the fringes of the campus. No hookups to the city sewer system were possible, and the developer would have had to build his own sewage treatment plant, which would discharge through the campus. Detective work uncovered the fact that the Fishery Department on campus had long-range plans for damming up the stream and using the lake for research. They developed the plan long before hearings on the developer's discharge permit were slated. During the discovery phase, the opposing counsel sarcastically referred to the "Pristine Creek" and "Mythical Lake," but the handwriting was on the wall. Treated sewerage would never be allowed to discharge into a fishery research lake. The developer quickly sold the land to the university.

The Expert needs to be able to place him or herself on the other side and view the case as an adversary. What if you were sitting behind the opposing counsel? What would your arguments be, your weaknesses and strengths? During a recent trial, as the other side was presenting its case, a great defense for them came to my mind that would have been hard for our side to counter. I spent that night attempting to figure out a solution—without success. The other side was never cognizant of that possible defense and the case eventually went in our favor. It was a close call. The opposing side had hired Experts at the last minute; their Experts did not have the time to uncover the subtle issues or the flexibility to switch technical theories.

Experts should attend all relevant depositions of both fact and Expert Witnesses. The Expert must pursue and understand the human aspects of a technical case. People make mistakes; science follows natural laws. The personality types and organization are as important to the Expert as they are to the lawyer. During a deposition I was critical of the way an industrial plant was operated by management. Later, I praised the same management for making an appropriate technical judgement that was not carried out. Thereupon the opposing attorney asked me if these people were geniuses or clowns; just what were they, given my opinions? I had attended the deposition of the key management person and noted that he was a strong technical person but had a dogmatic, gruff personality that made him a poor manager. Good technical people do not necessarily make good managers, which was my answer. This response would not have been possible if I had not attended this fact witness' deposition.

A good Expert will be able to develop the paradigms and metaphors for translating the technical information to a jury. Cases swing on what information is presented and how it is translated into the juror's minds. In a dispute involving the placement of a crude oil pipeline through a region containing a large aquifer used for diverting water for 1.5 million people, the original environmental assessment of a "worst case scenario" was an oil release based on a bulldozer accidentally rupturing the pipeline. This would cause minimal environmental impact. I restructured the worst case scenario: a terrorist attack on the pipeline, with thousands of barrels of oil released and significant poisoning of the aquifer. That paradigm shift in worst case perspective led to a compromise that justified rerouting the pipeline away from the aquifer.

A good Expert will brainstorm novel approaches to the technical issues. A proposed hotel/restaurant complex in an affluent residential neighborhood required some creative thought. The issue was to be fought on the application for treated sewage discharge into a nearby creek. The first line of defense was the contention that the creek flowed through a residential schoolyard, increasing the potential for children to come in contact with viruses in the water. The developer countered with a proposal for an advanced wastewater treatment plant that would produce "ultraclean" water. My creative retort was: would the developer be willing to use the effluent as the water in the hotel swimming pool? Of course not.

Trial exhibits are extremely important in technical cases, making seemingly esoteric statements come to life. Electrons need to be visualized. Groundwater contamination must be "seen" to be understood. Lawyers are often hesitant to invest significant expenditures for exhibits; this can be a big mistake. In the end, jurors exercise the right side of their brains (the conceptual) as well as the left side (the rational), and thought must be given as to how to fill both sides. A good Expert can break the technical down into simple, conceptual diagrams or models that get the point across. The team will be aided by a clear visualization of the more technical elements of the case. In one case, an industrial system, first modeled through the lawyer's instructions, was so poorly done that we considered suing the model builder for malpractice. Instead, it was scrapped for one that better illustrated the location.

A good Expert can help prepare the questions for direct and cross examination of witnesses. I sometimes like to fancy myself a law-yer and prepare the examination questions as if I were. That way, my attorney can take whatever questions they want, in elaboration of their own prepared questioning. I always have a big smile whenever one of "my" questions scores during examination. Sometimes I role play on a case with a lawyer friend with whom I have worked for fifteen years. We actually switch positions; I become the lawyer, he the Expert, as we explore a case. We also prepare witnesses by giving them my version of a direct or cross examination. It is usually less of a threat to the witness, since the questioning is more of a game, but the attorney can observe and explore the questions and responses for their preparation.

In summary: Hire the Expert early and utilize their full talents for the benefit of your case.

16

The Verdict

The verdict is in. The case is over.

But wait! The loser may appeal.

The judge may want to retry the case. Actually, the case is not over until both sides agree it's over. That can take years after the trial is over. Don't throw away any of the papers. Box up the documents and save them until your lawyer gives the final word.

This book is about over. However, I urge you to reread it before your next case. And here are some final thoughts I want to pass along.

PRACTICE

Practice, practice, practice. Nothing improves your ability to be an effective Expert Witness like practice. It is easy to revert back to your professional role after the case has ended. The comfort and pleasure of nonadversarial work becomes evident. You are again at peace with the world.

Unfortunately, the professional womb generally does not help you develop those important skills required to improve your status. You particularly need to practice your oral delivery skills. Consider ways to publicly put yourself on the line. In-volve yourself in community and political activities which require public exposure. For example, join an environmental organization involved in a cause of interest to you. Or join a professional association that is an advocate for one side of a cause. Public hearings may be scheduled where anyone with an interest in the matter can speak.

Hearings are a good forum for several reasons. First of all, they present an opportunity to make a prepared speech plus answer questions. Also, your expertise may have a direct bearing on the matter, so what you say has increased importance. As discussed in the chapter on marketing, you will likely be dealing with attorneys who hire Experts. Lastly, you are participating in the democratic process envisioned by our forefathers; this is how our system works.

You can also join a speakers' club like Toastmasters, dedicated to the improvement of the oral communications of its members. The two years I spent in Toastmasters were a strong impetus to me. The wide varieties of speeches I gave laid a foundation for adjusting to the legal system.

STUDY

Study the legal system in more depth. Make some free time to go down to the local courthouse and sit in on a trial. Try to figure out what is going on. A trial involving a small automobile accident has many of the same characteristics as a large product liability suit. Watch the jury as the witnesses are presented. Do your own critique. Make notes of the courtroom procedures you don't understand and strange words used, and then ask a lawyer friend to explain.

Keep track of newsworthy trials. Watch TV shows and movies involving courtroom action. Read books involving the legal system. Develop a sense for the lawyers and witnesses involved. What do the jurors say after the verdict is in? What events influenced their decisions?

Read about the history of the English legal system. The English system of Common Law is the basis of our system of justice. Jurors used to be the ones who witnessed the event or crime. Why did that system change? Did you know Expert Witnesses are a fairly new phenomenon?

You want to enter the courtroom with a feeling for the history and majesty of the system. And you need to feel comfortable with the surroundings. Simply stated, the more comfortable you are, the better you will do.

DETECTIVE WORK

You may have already discovered that the most demanding part of Expert witnessing is figuring out why some things happened in a case. What happened is obvious; why it happened is not. We professionals are good at developing correlations that technically explain the physical events in terms of science. We have a difficult time figuring out why human beings did what they did to cause an event to happen. Causality involves linking the scientific cause and effect with the psychological cause and effect. You must place yourself in the minds of others and play amateur psychologist.

Many times you will not be certain why somebody did something. You might develop alternative theories. Or you may never know.

Always keep in mind that physics is not on trial, people are. You need to link physical causes to the decisions of humans.

CONCLUSION

Expert witnessing demands a high level of creativity. You are at the vortex of the legal system, where physical and human conduct are examined in a particular way. If settlement is not achieved outside the courtroom, what a jury perceives in the microcosm of a trial becomes reality and finality. As the Expert, you are the cleanup hitter in the lineup. You neatly put the case together, tie up the loose ends, conveying its essence and meaning to the jury. You must draw on your background and on your power of persuasion to convince the jury. The ability to deliver is within; draw it out and use it.

17

The Future of Expert Witnessing

In many ways, the future looks bright. Brilliant minds are graduating from law schools in increasing numbers creating the potential for more litigation. The nature of litigation is more complex due to the reliance of society on technology. Governmental regulations in the areas of health, safety, and environment are more voluminous. All these trends mean that experts will be relied on with greater frequency. Indeed, expert witnessing is a growth industry.

These trends have resulted in a clogged court system weighted down by too many cases and drawn-out justice. One backlash is the strident calls for tort reform and the placement of strict limits on jury awards. No-fault insurance in some states has lowered the number of personal injury lawsuits. Other ways of settling litigation such as Alternative Dispute Resolution (ADR) are being advocated. These moves away from the formal legal system can diminish the need for experts.

Given these crosswise trends, I will attempt to make some predictions as to how expert witnessing will change and what it will take to be the expert for the 21st century.

HIGH TECH EXPERTS

High technology has been invading the domain of the expert in the areas of detective work and jury presentation. The latest techniques in chemical analysis are carted into the courtroom even before standardization and total acceptance by the professional organizations. Genetic testing of humans through DNA analysis is one such method in widespread use in the legal system while undergoing standardization in the profession. Novel theories of causation are flowing out of medical research. In toxic tort cases, the application of medical ecology theories based on damage to the human immune system by certain chemicals is controversial in the profession, but has been accepted in several court jurisdictions. It is based on the AIDS model for immune system damage where, instead of viruses attacking certain cells, chemicals do the damage. The field of forensic entomology is used to date the deaths of humans based on the life cycles of maggots that is quite different from its classical use to detect sources of insect infestation in civil cases. The revolutions in science will spin off directly to the experts who may be the ones fomenting change or those who quickly recognize uses in forensics. Experts who do not keep up in their fields with the latest technological advances and theories may find themselves unused and unwanted.

High tech in the courtroom as a means of communicating with the jury is undergoing rapid change. The traditional blackboard, slides, posters, and scale models are being supplemented with video and especially video animation. The incredible decrease in the cost of computer memory and speed in the form of powerful workstations has lowered the costs of animation by an order of magnitude in the last three years. Animations formerly costing one thousand dollars a second now run one hundred dollars a second. A riveting two-minute video animation shown to a jury demonstrating exactly what happened costs less than twenty thousand dollars and is affordable in cases in the million dollar range and up. With the television-watching generation showing up in the jury boxes, they will be treated more and more to the depiction of causation on the screen and in living color. Also, more courts are allowing video clips of alleged victims to be played to juries to show restrictions in their day-to-day activities as part of the damages claims. Experts will be actors in small dramas played out in front of cameras demonstrating various aspects of their cases. Today, in some cases, both parties agree to have their experts appear on tape and never show up in the courtroom. These forms of virtual reality will become more sophisticated. The experts with inclinations and abilities to capitalize on media will be in greater demand.

High tech approaches can backfire. In a recent case, the opposing expert and his attorney put together an amazing display on direct examination. They had poster boards, slides, video, scale model, well-choreographed chalkboard talk, and beautifully rehearsed questions and answers. I had never witnessed such a polished performance! When it was over they even seemed to bow to the jury with wide smiles before they sat down. Then came cross-examination, and the demeanor of the expert changed. He stuttered and sputtered. He became angry and sarcastic. His earlier performance with its high tech greatness contrasted so much from cross-examination that his credibility as an expert suffered.

High tech communication is a powerful tool when used appropriately. It should not dominate, but complement. As a postscript, the elegant scale model constructed by the opposition in the case just described was used by me so that their great effort was not entirely wasted.

JUNK SCIENCE

Novel theories are the substances of progress in science. Theories are proposed to explain physical phenomena in a different light. Scientists then work through the process of falsification to prove or disprove the theories. A theory, to hold up under scientific scrutiny, must not only explain incongruities in the current explanations, but must also not fail any test within its limits. If it does, the theory is falsified and repudiated. The entire process can take years, even decades. In the meantime, scientists embracing the latest novel theories will show up in the courtroom where the legal standards are less rigorous than the scientific standards. Will the judge allow a novel theory to be presented on the basis it is more probable than not? Or should the theory meet the rigorous standards of science for use by the expert? Or should the jury have to choose based on conflicting testimonies from experts with compelling scientific theories? The discretion of judges is sufficiently wide for the possibility of bizarre, offbeat, and even discredited theories to often make it into the courtroom.

Experts will be called upon with greater frequency to testify not only to the interpretation of data and facts but to the validity of the underlying theories. The basic tenets of science, how theories are formulated and proven, will undergo scrutiny. Novel theories will be promulgated by experts for a certain case or situation. It will be up to other experts to explain the fragility of new theories and the consequences of applying these theories without going through the normal channels of scientific scrutiny. Would the court system deal fairly with such theories as cold fusion, adaptive biology, or polywater? Do juries understand peer-reviewed work? What about the recent scandals involving contrived data reported in respectable journals to prove novel theories? Be prepared to discuss the culture of science as well as its application. Your own novel theory may be wonderful, but it will surely be attacked as being premature, not proven, and even wrong. The judge may even hire an expert to assist in determining whether it should see the light of day in the courtroom. Rest assured, you will be attached from the moment you offer the theory until the case is over.

The threat of junk science will always be present as long as there are experts willing to take the witness stand and promulgate offbeat theories. A small fraction of these theories may actually be accepted and eventually become the predominant view. However, the judge and jury need to know how the processes and culture of science work and how credible the theories are. The opposing expert is the educator of how science works. If your theory is novel or offbeat, be prepared for a withering cross-examination and great criticism from opposing experts.

FROM GLADIATOR TO NEGOTIATOR

One of the first cases I had as an expert involved contaminated farmland. A drilling company had disposed of drilling mud on the side of a hill. One week later a heavy rainstorm dislodged the mud and deposited it on adjacent farmland and a livestock pond. The farmer sued the drilling company for damages. Causation was not an issue, only damages. I was hired by both sides to figure out the extent of contamination. I looked hard, but could not find any evidence of long term, permanent damage. One day while taking samples, I asked the farmer how much he wanted for compensation. He had been thinking about it and gave me a number that I thought was reasonable. The defendants readily agreed, but the farmer's attorney did not — her contingency fee would be too low in her estimation. Further investigations verified no damage, and the case settled later on for half what the farmer wanted. My role in that case was not only as a researcher but that of a negotiator on unusual circumstances.

The legal system is based on the ancient gladiator model with lawyers and experts using paper and persuasion instead of swords and armor. One side wines and the other bores in the courtroom. A new wave of ways to resolve disputes is being tried under the catchall phrase, Alternative Dispute Resolution (ADR). Mediation and arbitration under less formal circumstances will allow parties to settle disputes with less acrimony and at lower cost. With the rising popularity of ADR, experts take on roles of detective and negotiator. They can work for a middle-ground instead of a polarization. They may even be on the panel that decides the disputes.

ADR is going to require the expert to develop broader skills. Looking for common grounds and ways to work out differences, experts will be freer to explain holes in their views in a less structured format. Hired guns will find their services not in demand, as the new breed of experts can work for the plaintiff or defendant with equal ease. More good experts will readily offer their services, where before they were turned off by the adversarial nature of the proceedings. Society will benefit from speedier justice and fairer resolution of disputes. Experts will survive and even prosper with a collegial system much closer to what they experience in their professional lives.

CONCLUSION

Expert witnesses play a strong role in the legal system. The role is changing under the accelerated technological advances, the latest scientific theories, and alternative forms of dispute resolution.

Appendix A

Directory of Organizations

Professional Forensic Societies

Defense Research Institute
Expert Witness Index
750 Lakeshore Drive
Chicago, IL 60611
(312) 944–0575

The Association of Trial
  Lawyers of America
1050 31st Street N.W.
Washington, DC 20007
(202) 965–3500

National Forensic Center
17 Temple Terrace
Lawrenceville, NJ 08648
(609) 883–0550

Association of Trial Behavior
  Consultants
Department of Speech
  Communications
University of Arizona
Tucson, AZ 85721
(602) 621–2211

The Society of Automotive
  Engineers
400 Commonwealth Drive
Warrendale, PA 15096
(412) 776–4841

American Bar Association
750 N. Lakeshore Drive
Chicago, IL 60611
(312) 944–0575

American Association of Cost
  Engineers
308 Monongahela Building
Morgantown, WV 26505
(304) 296–8444

National Academy of Forensic
  Engineers
National Society of
  Professional Engineers
1420 King Street
Alexandria, VA 22314
(703) 684–2880

Forensic Sciences Foundation,
   Inc.
225 S. Academy Boulevard
Colorado Springs, CO 80910
(303) 596–6006

Bar Associations

Alabama State Bar Association
415 Dexter Ave.
P.O. Box 671
Montgomery, AL 36101
(215) 269–7515

Alaska Bar Association
P.O. Box 279
Anchorage, AK 99510
(907) 272–7469

State Bar of Arizona
234 N. Central Ave., Suite 858
Phoenix, AR 85004
(205) 269–7515

Arkansas Bar Association
400 West Markham
Little Rock, AR 72201
(501) 375–4605

The State Bar of California
555 Franklin Street
San Francisco, CA 94102
(415) 561–8200

Colorado Bar Association
250 W. 14th Ave., Suite 800
Denver, CO 80204
(303) 629–6873

Connecticut Bar Association
101 Corporate Place
Rocky Hill, CT 06067
(203) 721–0025

Delaware State Bar
   Association
Carvel State Office Building
820 N. French Street
Wilmington, DE 19801
(302) 658–5278

District of Columbia Bar
1426 H Street NW, 8th Floor
Washington, DC 20005
(202) 638–1550

The Florida Bar
650 Appalachee Parkway
Tallahassee, FL 32399
(904) 222–5286

State Bar of Georgia
84 Peachtree St., 11th Floor
Atlanta, GA 30303
(404) 522–6255

Hawaii State Bar
P.O. Box 26
Honolulu, HI 96810
(808) 537–1868

Idaho State Bar
P.O. Box 895
Boise, ID 83701
(208) 342–8958

Illinois State Bar Association
Illinois Bar Center
Springfield, IL 62701
(217) 525–1760

Indiana State Bar Association
230 S. Ohio Street
Indianapolis, IN 46204
(317) 639–5465

Iowa State Bar Association
1101 Fleming Building
Des Moines, IA 50309
(515) 243–3179

Kansas Bar Association
P.O. Box 1037
Topeka, KS 66601
(913) 234–5696

Kentucky Bar Association
W. Main at Kentucky River
Frankfort, KY 40601
(502) 564–3795

Louisiana State Bar
   Association
210 O'Keefe Ave., Suite 600
New Orleans, LA 70112
(504) 566–1600

Maine State Bar Association
P.O. Box 788
Augusta, ME 04330
(207) 622–7523

Maryland State Bar
   Association
207 E. Redwood, Suite 905
Baltimore, MD 21202
(301) 685–7878

Massachusetts State Bar
   Association
One Center Plaza
Boston, MA 02108
(617) 523–4529

State Bar of Michigan
306 Townsend
Lansing, MI 48933
(517) 372–9030

Minnesota State Bar
   Association
430 Marquette Ave., Suite 403
Minneapolis, MN 55402
(612) 333–1183

Mississippi State Bar
Association
620 N. State St.
P.O. Box 2168
Jackson, MS 39205
(609) 948–4471

Missouri Bar
326 Monroe
P.O. Box 119
Jefferson City, MO 65102
(314) 635–4128

State Bar of Montana
P.O. Box 4669
Helena, MT 59601
(406) 442–7660

Nebraska State Bar
Association
206 S. 13th Street
Lincoln, NE 68508
(402) 475–7091

State Bar of Nevada
834 Willow Street
Reno, NV 89502
(702) 329–4100

New Hampshire Bar
Association
1850 Elm St.
Box 719
Manchester, NH 03305
(603) 669–1000

New Jersey State Bar
Association
172 W. State Street
Trenton, NJ 08608
(609) 394–1101

State Bar of New Mexico
P.O. Box 25883
Albuquerque, NM 87125
(505) 842–6132

New York State Bar
Association
One Elk Street
Albany, NY 12207
(518) 463–3200

North Carolina State Bar
208 Fayetteville St. Mall
P.O. Box 25850
Raleigh, NC 27611
(919) 828–4620

State Bar Association of North
Dakota
P.O. Box 2136
Bismarck, ND 58502
(701) 255–1404

Ohio State Bar Association
33 West 11th Avenue
Columbus, OH 43201
(614) 421–2121

Oklahoma Bar Association
19091 N. Lincoln Boulevard
Oklahoma City, OK 73152
(405) 524–2365

Oregon State Bar Association
1776 Southwest Madison St.
Portland, OR 97205
(503) 224–4280

Pennsylvania Bar Association
P.O. Box 186
Harrisburg, PA 17108
(717) 238–6715

Rhode Island Bar Association
1804 Fleet National Bank Bldg.
Providence, RI 02903
(401) 421–5740

South Carolina Bar
P.O. Box 11039
Columbia, SC 29211
(803) 799–6653

State Bar of South Dakota
222 E. Capitol
Pierre, SD 57501
(605) 224–7554

ennessee Bar Association
622 West End Avenue
ashville, TN 37205
15) 383–7421

The State Bar of Texas
1414 Colorado
Austin, TX 78711
(512) 463–1463

Utah State Bar
425 East First South
Salt Lake City, UT 84111
(801) 531–9077

Vermont Bar Association
P.O. Box 100
Montpelier, VT 05602
(802) 223–2020

Virginia State Bar
700 East Main, Suite 1622
Richmond, VA 23219
(804) 786–2061

Washington State Bar
    Association
505 Madison Street
Seattle, WA 98104
(206) 622–6054

West Virginia Bar Association
P.O. Box 346
Charleston, WV 25322
(304) 342–1474

State Bar of Wisconsin
402 West Wilson Street
Madison, WI 53702
(608) 257–3838

Wyoming State Bar
16th & Capitol
P.O. Box 109
Cheyenne, WY 82003
(307) 632–9061

Appendix B

Recommended Practices for Design Professionals Engaged as Experts in the Resolution of Construction Industry Disputes*

Preamble

Experts are vitally important to contemporary American jurisprudence. They review and evaluate complex technical issues and explain their findings and opinions to lay triers of fact for the latter's consideration in reaching a verdict.

Experts retained by opposing parties may disagree. In all instances, such disagreements should emanate only from differences in professional judgment.

These recommendations have been developed from the belief that adherence to them will help experts provide to triers of fact substantiated professional opinions unbiased by the adversarial nature of most dispute resolution proceedings. The

---

*Reprinted with permission from ASFE: The Association of Engineering Firms Practicing in the Geosciences. For further information, contact ASFE: The Association of Engineering Firms Practicing in the Geosciences, 8811 Colesville Road, Suite G-106, Silver Spring, MD, 20910, (301) 565-2733.

organizations which endorse these recommendations do not require any individual to follow them.

Recommendations

It is the obligation of an expert to perform in a professional manner and serve without bias. Toward these ends:

1. **The expert should avoid conflicts of interest and the appearance of conflicts of interest.**

   Commentary: Regardless of the expert's objectivity, the expert's opinion may be discounted if it is found that the expert has or had a relationship with another party which consciously or even subconsciously could have biased the expert's services or opinions. To avoid this situation, experts should identify the organizations and individuals involved in the matter at issue, and determine if they or any of their associates have or ever had a relationship with any of the organizations or individuals involved. Experts should reveal any such relationships to their clients and/or clients' attorneys to permit them to determine whether or not the relationships could be construed as creating or giving the appearance of creating conflicts of interest.

2. **The expert should undertake an engagement only when qualified to do so, and should rely upon other qualified parties for assistance in matters which are beyond the expert's area of expertise.**

   Commentary: Experts should know their limitations and should report their need for qualified assistance when the matters at issue call for expertise or experience they do not possess. In such instances, it is appropriate for experts to identify others who possess the required expertise, and to work with them. Should an expert be asked to exceed his or her limitations and thereafter be denied access to other professionals, and should the expert be requested to continue association with the case, the expert should establish which matters he or she will and will not pursue; failing that, the expert should terminate the engagement.

3. **The expert should consider other practitioners' opinions relative to the principles associated with the matter at issue.**

    Commentary: In forming their opinions, experts should consider relevant literature in the field and the opinions of other professionals when such are available. Experts who disagree with the opinion of other professionals should be prepared to explain to the trier of fact the differences which exist and why a particular opinion should prevail.

4. **The expert should obtain available information relative to the events in question in order to minimize reliance on assumptions, and should be prepared to explain any assumptions to the trier of fact.**

    Commentary: The expert should review those documents, such as tenders and agreements, which identify the services in question and any restrictions or limitations which may have applied. Other significant information may include codes, standards and regulations affecting the matters in dispute, and information obtained through discovery procedures. If pertinent to the assignment, the expert should also visit the site of the event involved and consider information obtained from witnesses. Whenever an expert relies on assumptions, each assumption should be identified and evaluated. When an assumption is selected to the exclusion of others, the expert should be able to explain the basis for the selection.

5. **The expert should evaluate reasonable explanations of causes and effects.**

   Commentary: As necessary, experts should study and evaluate different explanations of causes and effects. Experts should not limit their inquiry for the purpose of proving the contentions advanced by those who have retained them.

6. **The expert should strive to assure the integrity of tests and investigations conducted as part of the expert's services.**

   Commentary: Experts should conduct tests and investigations personally, or should direct their performance through qualified individuals who should be capable of serving as expert or factual witnesses with regard to the work they performed.

7. **The expert witness should testify about professional standards of care** only with knowledge of those standards which prevailed at the time in question, based upon reasonable inquiry.**

   Commentary: When a design professional is accused of negligence, the trier of fact must determine whether or not the professional breached the applicable standard of care. A determination of the standard of care prevailing at the time in question may be made through investigation, such as the review of reports, records, or opinions of other professionals performing the same or similar services at the time in question. Expert witnesses should identify standards of care independent of their own preferences, and should not apply present standards to past events.

**Standard of care is commonly defined as that level of skill and competence ordinarily and contemporaneously demonstrated by professionals of the same discipline practicing in the same locale and faced with the same or similar facts and circumstances.

8. **The expert witness should use only those illustrative devices or presentations which simplify or clarify an issue.**

    Commentary: The attorney who will call the expert as a witness will want to review and approve illustrative devices or presentations before they are offered during testimony. All illustrative devices or presentations developed by or for an expert should demonstrate relevant principles without bias.

9. **The expert should maintain custody and control over whatever materials are entrusted to the expert's care.**

    Commentary: The preservation of evidence and the documentation of its custody and care may be necessary for its admissibility in dispute resolution proceedings. Appropriate precautions may in some cases include provision of environmentally controlled storage.

10. **The expert should respect confidentiality about an assignment.**

    Commentary: All matters discussed by and between experts, their clients and/or clients' attorneys should be regarded as privileged and confidential. The contents of such discussions should not be disclosed voluntarily by an expert to any other party, except with the consent of the party who retained the expert.

11. **The expert should refuse or terminate involvement in an engagement when fee is used in an attempt to compromise the expert's judgment.**

    Commentary: Experts are employed to clarify technical issues with objectivity and integrity. Experts should either refuse or terminate service when they know or have reason to believe they will be rewarded for compromising their objectivity or integrity.

12.   **The expert should refuse or terminate involvement in an engagement when the expert is not permitted to perform the investigation which the expert believes is necessary to render an opinion with a reasonable degree of certainty.**

Commentary: It is the responsibility of experts to inform their clients and/or their clients' attorneys about the scope and nature of the investigation required to reach opinions with a reasonable degree of certainty, and the effect which at any time, budgetary or other limitations may have. Experts should not accept or continue an engagement if limitations will prevent them from testifying with a reasonable degree of certainty.

13.   **The expert witness should strive to maintain a professional demeanor and be dispassionate at all times.**

Commentary: Particularly when rendering testimony or during cross-examination, expert witnesses should refrain from conducting themselves as though their service is a contest between themselves and some other party.

Endorsing Organizations and Date of Endorsement*

ASFE/The Association of Engineering Firms Practicing in the Geosciences—February 15, 1988

American Academy of Environmental Engineers—March 15, 1988

American Association of Cost Engineers—July 9, 1988

American Congress on Surveying and Mapping—January 24, 1989

---

*As of April 24, 1989.

American Consulting Engineers Council—January 18, 1988

American Council of Independent Laboratories—April 8, 1988

American Institute of Architects—March 14, 1988

American Institute of Certified Planners—April 25, 1987

American Nuclear Society—November 2, 1988

American Public Works Association—April 4, 1989

American Society of Agricultural Engineers—October 20, 1988

American Society of Civil Engineers—October 23, 1988

American Society of Consulting Planners—April 30, 1988

American Society of Landscape Architects—August 15, 1987

American Society of Safety Engineers—June 1, 1988

Association of Energy Engineers—November 4, 1988

Association of Engineering Geologists—April 23, 1988

California Geotechnical Engineers Association—June 23, 1988

Illuminating Engineering Society of North America—June 7, 1988

Interprofessional Council on Environmental Design—March 17, 1988

National Academy of Forensic Engineers—January 26, 1988

National Society of Professional Engineers—January 21, 1988

Structural Engineers Association of Illinois—December 6, 1988

Washington Area Council of Engineering Laboratories—June 23, 1988

Appendix C

References

Dombroff, Mark A. "Prepare and Present Your Expert Witness," *For the Defense* (August 1984), pp. 15–23.

Humphreys, Hugh C. "Cross-Examining the Expert; Some Tips from the Bench," *Trial* (October 1987) pp. 75–78.

Jones, Richard T. "Impeaching Your Opponent's Expert," *The Brief* (Summer 1985), pp. 42–45.

Klein, Stanley J., P.E., C.E.C. "Making the Most of Your Expert," *Connecticut Bar Journal.* 46:483–495 (1972).

Krueger, Eric F., P.E. "The Engineering Expert," *Advocates Quarterly*, 1983, pp. 326–331.

Nunnally, Knox D. "Use of Experts," State Bar of Texas Professional Development Program, Advanced Personal Injury Law Course (July 1987), pp. U1-U47.

Reynolds, Ernest (Skip) III. "The Selection and Use of the Defense Expert," State Bar of Texas Professional Development Program, Experts in Litigation: A Performance Enhancement Course, Fort Worth, TX (Autumn 1987), pp. C1-C64.

Ring, Leonard M. "Choosing and Presenting Your Expert," *The Brief* (Summer 1985), pp. 35–41.

Rossini, Dino, and Cantilli, Edmund J. "Using Engineering Experts Effectively," *Case & Comment* (March-April 1982), pp. 40–44.

Ryan, Joseph, Jr. "Making the Plaintiff's Expert Yours," *For the Defense* (November 1982), pp. 12–20.

Ryder, Frederick L. "Engineering in Litigation; Some Misconceptions," *Trial* (June 1985), pp. 64–66.

APPENDIX D

Letter to Attorney and Sample Resume

Dear Attorney:

I am interested in being an expert witness in litigation. My areas of expertise include:

- Refinery and Petrochemical Plant Construction
- Industrial Safety and Maintenance
- Accident Reconstruction

My credentials are a BS in Mechanical Engineering and 32 years of experience in the refining and petrochemical industry. I recently took early retirement from Petro Industries where I was the Maintenance Superintendent.

Please call me at 713: XXX-XXXX if you have a case in my areas of expertise.

Sincerely,

Ronald Cowart, P.E.

Encl: Resume

Ronald W. Cowart, P.E.
10919 Braes Forest
Houston, Texas 77071
713: XXX-XXXX

Areas of Expertise
> Refining and Petrochemical Plant Construction
> Industrial Safety and Maintenance
> Accident Reconstruction

Education
> BS Mechanical Engineering, 1960, Kansas
> OSHA Safety Short Course, 1980
> Management Training Institute, 1984
> Haswopper Course (hazardous waste) 1990

Professional Experience
> 1960–1964, Bechtel Constructors. Planning coordinator and construction supervisor on ethylene chemical plant in Louisiana.
>
> 1965–1978, Humble Chemicals. Startup engineer and operating supervisor at ethylene plant.
>
> 1978–1992, Petro Industries. Superintendent in charge of all maintenance activities at an integrated refinery, petrochemical plant. Supervised plant-worker accident investigations and was in charge of plant safety and compliance with OSHA regulations.

Related Activities
> Professional Engineer in Louisiana and Texas
> Member of American Society of Mechanical Engineers
> Short Course on Expert Witnessing, 1992
> Deposed in company-related lawsuit, 1988

Index